ARMS RACE THEORY

Recent Titles in
Contributions in Military Studies

Behind a Curtain of Silence: Japanese in Soviet Custody, 1945-1956
William F. Nimmo

Armed Forces on a Northern Frontier: The Military
in Alaska's History, 1867-1987
Jonathan M. Nielson

The Aegean Mission: Allied Operations in the Dodecanese, 1943
Jeffrey Holland

The World Turned Upside Down: The American Victory
in the War of Independence
John Ferling, editor

U.S. Unilateral Arms Control Initiatives: When Do They Work?
William Rose

Transfer of Arms, Leverage, and Peace in the Middle East
Nitza Nachmias

Missile Defenses and Western European Security: NATO Strategy,
Arms Control, and Deterrence
Robert M. Soofer

The U.S. War Crimes Trial Program in Germany, 1946-1955
Frank M. Buscher

Democracy under Siege: New Military Power in Latin America
Augusto Varas, editor

Not Shooting and Not Crying: Psychological Inquiry into Moral Disobedience
Ruth Linn

"Seeing the Elephant": Raw Recruits at the Battle of Shiloh
Joseph Allan Frank and George A. Reaves

Civilian Indoctrination and the Military: World War I and Future
Implications for the Military-Industrial Complex
Penn Borden

ARMS RACE THEORY

Strategy and Structure of Behavior

Craig Etcheson

Contributions in Military Studies, Number 87

COLIN GRAY, SERIES ADVISER

Greenwood Press

New York • Westport, Connecticut • London

Library of Congress Cataloging-in-Publication Data

Etcheson, Craig, 1955-
Arms race theory : strategy and structure of behavior / Craig Etcheson.
p. cm.—(Contributions in military studies, ISSN 0883-6884 ; no. 87)
Bibliography: p.
Includes index.
ISBN 0-313-26254-3 (lib. bdg. : alk. paper)
1. Arms race. 2. Arms race—Mathematical models. I. Title. II. Series.
UA10.E85 1989
355'.02—dc19 88-38995

British Library Cataloguing in Publication Data is available.

Library of Congress Catalog Card Number: 88-38995
ISBN: 0-313-26254-3
ISSN: 0883-6884

First published in 1989

Greenwood Press, Inc.
88 Post Road West, Westport, Connecticut 06881

Printed in the United States of America

The paper used in this book complies with the Permanent Paper Standard issued by the National Information Standards Organization (Z39.48-1984).

10 9 8 7 6 5 4 3 2 1

This work is dedicated
to all of my teachers,
especially the two most important ones,
my mother and father.

CONTENTS

ILLUSTRATIONS

Figures

Tables

ACKNOWLEDGMENTS

In the more than a decade this book has been in preparation, literally scores of teachers, colleagues and students have influenced and aided my research. Many of the hundreds of scholars whose work I discuss in this book have generously criticized and assisted various stages of my research. Dozens of librarians, computer technicians, and others have labored selflessly. **So many** have helped. To list all these people would be too tedious; they know who they are: Thank you all.

Still, I would be remiss if I failed to acknowledge the special debts owed to a special few. James N. Rosenau, Richard K. Ashley, Dwain Mefford, and Walt Scacchi, the members of my graduate committee at the University of Southern California, patiently sustained their criticism over many years. Dina Zinnes and Phil Schrodt, more distant but nonetheless persistent sources of criticism and inspiration, were also of more help than they know. A casual comment about puzzles by Stephen Majeski in Atlanta in the spring of 1984 stimulated me to draft the first three chapters in an eleven day period. Criticism by Stuart Thorson, J. David Singer, Rein Taagepera, and several anonymous reviewers proved valuable late in the manuscript preparation process.

Perhaps the greatest tribute that I can give to these scholars is to note that, even when they disagree with some of my conclusions (as they all do, to one extent or another), their commitment to the process of scholarly inquiry propelled them to make valuable and sustained contributions to my creative spark, such as it is. This gives me confidence that the spirit of criticism evidenced in their commitment will continue to inform progress in our understanding of strategy and the structure of behavior in arms races.

If we program a machine for winning a war, we must think well what we mean by winning. A learning machine must be programmed by experience. The only experience of a nuclear war which is not immediately catastrophic is the experience of a war game. If we are to use this experience as a guide for our procedure in a real emergency, the values of winning which we have employed in the programming games must be the same values which we hold at heart in the actual outcome of a war. We can fail in this only at our immediate, utter, and irretrievable peril. We cannot expect the machine to follow us in those prejudices and emotional compromises by which we enable ourselves to call destruction by the name of victory.

-- Norbert Weiner, 1961

ARMS RACE THEORY

INTRODUCTION: STRATEGY AND THE STRUCTURE OF BEHAVIOR IN ARMS RACES

A generation has now come to maturity, the first in history to live all their days in the knowledge that delivered into human hands is that which was formerly the sole property of Shiva and Yahweh: the power to destroy worlds. The weight of this terrible power threatens to drag down not merely the entire generation, but future generations as well. Wondering how we got here, and where we can go from here, led to this study. Yet this effort quickly became a "meta-study" on how these questions have been addressed by scholars and the implications of their orientations for the resulting knowledge claims. This is so because any such inquiry cannot escape the overriding tension between ideas and action, a critical tension permeating both subject and object.

Social scientists have long scrutinized problems of armaments accumulation, and thus this inquiry necessarily begins with a re-search of the existing work. Aspects of the theories and the methods employed in the study of arms accumulation will demand particular attention. I shall suggest, however, that it is the interface between theory and method that lies at the crux of the matter. This study explores new theories, new methods, and, most significantly, a new conception of the relationship between the two in the study of arms accumulation.

Traditional social scientific practice in the behavioral schools has encouraged the neglect of concepts of meaning, purpose, and teleology, a norm at odds with the fact of the primacy of values in social behavior. Increasingly, however, as these concepts have infiltrated by virtue of their necessity back into the lexicon of legitimate social scientific inquiry, analysts of politico-military affairs are beginning to avail themselves these new opportunities for understanding. Values are beginning to return to their rightful place at the center of social inquiry. And values, in the guise of plans and strategies, lie at the conceptual center of the present effort. It is now becoming possible to integrate these concepts into our models and simulations, and here we make first explorations of this new territory.

Admittedly this is a tall order. As Plato has suggested, a good way to begin such an ambitious enterprise is by, first, "the taking in of scattered particulars under one Idea, so that everyone understands what is being talked about."[1] Our idea has to do with the problem of interaction in armaments accumulation.

Note

1. Christopher Alexander, Notes on the Synthesis of Form (Cambridge, MA: Harvard University Press, 1964), p. viii; cf. Plato, Phaedrus, Verse 265D.

1 THE PROBLEM OF INTERACTION IN ARMS ACCUMULATION

Do military bureaucracies interact? Certainly they act. Are these actions patterned and intentional? Do they constitute an interactive system? If so, in what ways?

Anecdotal evidence and common-sense interpretation of everyday events strongly suggests to most people that large modern military bureaucracies do indeed interact with one another. Military establishments would seem to interact over a wide range of relations, from motivation and emotion to perception and memory through planning and projections and finally in deployments and combat. Scholars studying this phenomenon, however, have been unable to establish a clear and consistent empirical link that would tend to substantiate the theory that nations interact in so-called arms races. The persistent finding has been that there is little or no evidence of interaction between nations that are engaged in apparently competitive or interactive armaments accumulation behavior. This result is particularly pronounced in studies of military interaction between the United States and the Soviet Union. The finding is that there is apparently no military interaction between the world's two dominant military powers. Such a conclusion is genuinely puzzling.

1.1 Interaction Constitutes a Genuine Puzzle

A genuine puzzle in the study of international relations is said to arise when we observe a recurrent patterned sequence of events or outcomes and find our curiosity piqued with the realization that the processes giving rise to the patterns are neither self-evident nor easily explained.[1] Some apparently persistent pattern of events *seems* like it should fit with another pattern. Yet despite a concerted methodological assault (and sometimes perhaps because of it), the dependent variables we seek to understand may resist explanation, yielding instead counter-intuitive or nonsensical findings. The result can be genuine puzzlement.[2]

Many of those studying patterns of armaments accumulation have achieved genuine puzzlement and rightly so, for there are puzzles here. For decades it was a matter of almost casual presumption that interaction is a central element in armaments races and war. In his monumental tome, *A Study of War*, Quincy Wright proposed that when "armaments budgets, personnel, and material are rising at an accelerating rate, it may be assumed that international tensions are increasing and that states are morally, as well as materially, rearming. Armaments races, evidenced by such statistics, constitute a form of international relations closely related to war and often ending in war itself."[3] Or so one would think. Intuition tells us that when two countries are directing hostile behavior at one another and are accumulating stocks of military capability, an interaction of a most interesting and profound order is taking place. We anticipate that the pattern of armaments accumulation for the United States, for example, would somehow "map" into or fit the pattern of the Soviet Union. Yet analysts have found that such patterns generally do not seem to fit well. As one leading student of arms races has expressed it, we "believe that nations do interact, yet a rather extensive set of data and sophisticated statistical analyses say they do not. Thus puzzlement arises. . . . "[4]

The evidence Dina Zinnes mentions is extensive and persuasive. Evidence that arms accumulation cannot be explained as an interactive process has been amassed by a formidable array of scholars.[5] The weight of this evidence persuaded many to abandon the idea of interaction in favor of what might be characterized as variants on a Schumpeterian perspective, where arms accumulation is seen as a consequence of an objectless disposition on the part of military bureaucracies to unlimited expansion.[6] This has resulted in several arms studies focusing on the role of factors internal to the nation-state.[7] Yet these studies did not solve the puzzle.

Studies of arms accumulation have come full circle. In general the findings of studies on internal factors in arms races are no less problematic than those exploring interactive behavior. This has led investigators back to the notion that, while internal variables are important, the idea of interaction should be re-examined rather than rejected.[8] This is where I propose to begin.

1.2 Interaction Is an Ambiguous Theory

Theories of interaction in arms races have problems as social scientific concepts. Indeed, the very idea of an arms race is fraught with ambiguity and is of uncertain etymological origin. Legend has it that circa 1912 the imaginative editor of a London tabloid coined the phrase "arms race" to characterize Anglo-German naval competition. The term has since captured the fancy of persons public and private, lay and academic. Uncritical acceptance of this term, and its consequent extension by analogy, contributes many unfortunate images and assumptions to public and academic debate on the topic. While "arms race" is a lovely and enchanting metaphor, it is a poor analogy in that it implies a parallel between the form of a sporting event and the form of strategic arsenal accumulation. "A comparison," as one pioneer aptly observed, "of the two kinds of races will reveal their psychological unlikeness."[9]

The two forms are not isomorphic in a number of critical respects. Lewis Fry Richardson argues, for example, that the relationship between acceleration and velocity is negative in a foot race but positive in an arms race. "Arms race" implies that the contestants both have the same goal, operate under the same rules of play (to which both willingly adhere), and that they will be running over an identical course. None of these implications would appear to be particularly useful for gaining a deeper understanding of the phenomenon of military interaction. They may in fact contribute nothing more than additional confusion, for many studies fail even to approach "the fundamental question whether there has been an arms race at all."[10] "In spite of wide use," another writer concludes, "the concept reflects little empirically based understanding of the processes it purports to encompass."[11]

More seriously, use of the term *arms race* would seem to prejudge the matter. In some cases, military relations between states exhibit no obvious pattern of covariation, and here the concept of an arms race would appear to be of questionable relevance. An arms race suggests that two or more parties are somehow "racing forward" together. In the case of the premier arms race, that between the United States and the Soviet Union, some analysts argue that this image is so fundamentally misleading as to constitute little more than an ideological polemic.[12]

The term *armaments accumulation* seems to present fewer definitional problems. It is not so loaded with perjorative and prescriptive baggage as is *arms race*. It is more neutral in a denotative sense, with respect to the magnitude, direction, and causality of the relationship. A single weapon constitutes an accumulation, regardless of whether internal or external factors were instrumental in its genesis. Unlike arms race, the term *arms accumulation* encompasses interaction without presuming it. Moreover, *accumulation* can easily accommodate both negative and positive change; while the notion of "negative accumulation" is intuitively clearcut, the idea of a negative race seems unnecessarily oxymoronic. For these reasons, the term *arms accumulation* will be favored in this analysis in the place of the more traditional term, *arms race*.

What, then, is meant by a theory of interaction in arms races or armaments accumulation? A scientist named Lewis Fry Richardson, inspired by his experiences as an ambulance driver in France during World War I, devoted considerable thought to precisely this question. Richardson grounded his ideas about arms races with a *ceteris paribus* clause: ". . . the scope of the present theory is restricted to the interaction of groups which style themselves powers, which are proud of their so-called sovereignty and independence, are proud of their armed might, and are not exhausted by combat. This theory is not about victory and defeat."[13]

Richardson thus summarized the theoretical kernel which informs his mathematical models of arms races. He went on to express the theorized processes of armaments accumulation in terms of a pair of differential equations, the classic action-reaction model.[14] "The processes described by the ensuing equations . . . is what would occur if instinct and tradition were allowed to act uncontrolled":

$$dx/dt = ky - ax + g \qquad [1.1]$$

$$dy/dt = lx - by + h \qquad [1.2]$$

As Michael Squires points out, the "action-reaction model should not be overestimated as a theory."[15] In fact, the model is not a theory at all; it is a model. It is important to be clear about the distinction between the theory and the model put forward by Richardson. Theories are ideas about relationships between entities; models are symbolic representations of theories, sometimes in testable form. Richardson's theory posited that, because statesmen are motivated by instincts of fear and aggression and by traditions of *Realpolitik* and *raison d'etat*, armaments accumulation tends to proceed in a predictable, indeed, in a mechanical fashion. His action-reaction model, on the other hand, was an attempt to represent this theory in an explicit mathematical form. "The" model is but an instance of the theory.

We agree with Stephen Majeski[16] and others that Richardson's principal contribution to the tradition he founded was not his specific models but rather the theory kernel, combined with the spirit of critical scientific inquiry. The verbal, non-formal and allegorical nature of his theoretical presentation admits to many possible representations. This is affirmed in the considerable variety of models posited by Richardson and the explosion of models in the literature produced by his students. Even so, however, his students have been reluctant to stray too far from the original formulations.

The original formulation is based upon an awareness that the *sovereign state* recognizes no other as equal. Frequently, neither do the agents of the state recognize others as equal. Political theorists have long recognized this aspect of the state and its agents. G.W.F. Hegel perhaps said it best: "the substance of the state" is "its absolute power against everything individual and particular, against life, property and their rights, even against societies and associations."[17] The idea of sovereignty originated with the prince and even as transferred to the state the concept refers to supreme, absolute, and undisputed ascendancy. The state recognizes no other as equal. This supposedly legitimizes the use of force by the state. The state claims and asserts a monopoly on legitimate violence internally but must because of the "state of nature" share the right externally in the international system.

To execute external force, modern states establish agencies dedicated to the promulgation of organized violence: the ministry of defense. The nature of such an organization has been well stated by Max Weber: "the purely bureaucratic type of administrative organization [is] . . . the most rational known means of exercising authority over human beings."[18] And the principal means by which the authority of the bureaucracy is brought to bear is *rationality*. According to Weber, bureaucracy is the supreme form of domination precisely because of its relentless application of rationality to administration. "The ruled, for their part, cannot dispense with or replace the bureaucratic apparatus once it exists, for it rests upon expert training, a functional specialization of work, and an attitude set on habitual virtuosity in the mastery of single yet methodically integrated functions."

The central focus here is on calculation: the cold, objective "discharge of business." "'Objective' discharge of business primarily means a discharge of business according to *calculable rules* and 'without regard for persons.'" The greater the degree of elimination of the human element from the business of bureaucratic administration, the more perfect is the rationality. "Bureaucracy develops the more perfectly, the more it is 'dehumanized,' the more it completely succeeds in eliminating from official business love, hatred, and all purely personal, irrational and emotional elements which escape calculation." The ideal type of bureaucracy functions just like a machine. Indeed, it *is* a machine, a perfect input-output device. "Bureaucracy is *the* means of transforming social action into rationally organized action."

In modern war, this is intensified. "War in our time is a war of machines, and this makes central provisioning technically necessary . . . ." Weber saw the survival of the state and the bureaucracy in their modern forms as being two aspects of the same whole: "the sure instincts of the bureaucracy for the conditions of maintaining its *own* power in the home state (and through it, in opposition to other states) are inseparably fused with this canonization of the abstract and 'objective' idea of 'reasons of state.'" The ministry of defense, dedicated to enforcement of the state's impulses to domination, therefore occupies a special place among the agencies comprising the state and the system of states.

National policy thus crystallizes into what Thomas Schelling has referred to as the "diplomacy of violence."[19] State policy organs are outfitted for the routine maintenance of a strategic military threat to the external environment. Bargaining on a strategic military plane is then reduced to a politically sterile process of technological management, constituting at most a very vulgar form of strategy. It is not too much to say that under these conditions, strategy is abdicated in favor of mere instrumental operations.[20] This is crucial to the logic of arms races and to a critique of arms accumulation modeling. Political bargaining requires the reflective recognition and empathetic awareness of an Other. Technological management, by contrast, implies the complete objectification of the environment such that there are no Others, only Self and things to be operated upon. Armaments accumulation processes appear mechanical, then, because the political dynamics of social reality are replaced with a static technological management--in our conceptions, our practices, *and* our theories.

1.3 Much Depends upon How Interaction Is Conceived

Much of our understanding of interaction in arms accumulation depends on how that interaction is conceptualized. In a pathbreaking study of Richardson models, Squires concludes his discussion of the relative merits of internal and external models of armaments accumulation with the observation that "neither theory has been totally satisfactory as an explanation of the formation of defense budgets of the United States and the U.S.S.R.; as these budgets, and the arms they represent, are major elements of the process of U.S.-U.S.S.R. interaction, we have yet to come to a satisfactory and complete theory of that interaction."[21] This conclusion displays one of the central features of most existing studies of interaction in armaments accumulation, that is, the representation of arms by defense budgets. Figure 1.1 depicts this feature. The aim in most empirical studies of interaction in armaments accumulation has been to test the hypothesized relationship "B" in Figure 1.1 as a surrogate for or operationalization of the theorized relationship "A." The general idea of interaction tested here has been that when the arms budget of the other is threatening, one will increase one's own military expenditure in some fixed proportion to that threat, say 1.1 to their 1.0. This is not a particularly sophisticated formulation, but one recommended by parsimony. When the "B" relation is not found, this is often interpreted as evidence that there is no "A" relation.

Figure 1.1

Budget as Surrogate for Theorized Variable

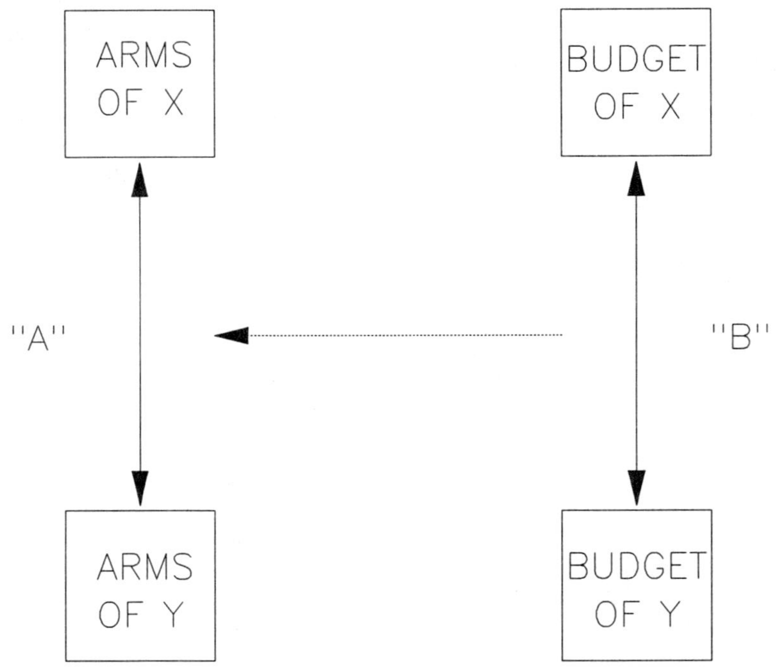

Unfortunately, it seems strongly likely that armaments accumulation processes are a bit more complicated than this. Leading students of the Richardson School have noted that the methods used to arrive at this conclusion, while reducing the problem to manageable proportions, also abstract out the inherent complexity underlying this social phenomenon.

A large part of the problem lies in what I will call the "ghost of mechanism." Models cannot be spontaneous in the same way that people can. People can be unpredictable at times, as can the social systems to which they belong. But models cannot. This is what I mean when I say that these models are haunted by the ghost of mechanism. We will come back to this idea again and again in the course of this study, for it is a central problem of social science that our models fail to embody purposiveness. This is the problem of intentionality.

The last four major research undertakings in the Richardson Tradition[22] have all concentrated on the development of more sophisticated perspectives on armaments accumulation, while always operating securely within the traditional methodological perspective. These efforts have avoided becoming mired in the philosophical problems associated with intentionality and the production of subjectivity. How does intentionality arise? Whence derives purpose? The traditional approach in the Richardson Tradition has been to deal with this challenge by ignoring it, putting the effort into observation of external relations of physical correlation, as between arms budgets. Concepts of meaning, purpose, and intentionality are integrated as afterthoughts, if at all.

In some respects it is difficult to understand why progress has been so halting in the Richardson Tradition. The crucial advances in systems theory necessary to the specification of more adequate models of arms accumulation phenomena occurred more than thirty years ago, but have not been exploited in the Richardson Tradition. W. Ross Ashby elaborated the mechanisms of "ultra-stability" required to conceptualize adaptation in the behavior of organisms.[23] Ashby argued that the notion of stability is inadequate to explain the persistence of organisms. A stable system or organism adjusts its behavior sufficiently to survive fluctuating conditions in the environment. This same notion of stability is seen in existing optimal control models of arms accumulation.

Mechanisms of simple stability, however, fail to account for the varieties of adaptive behavior seen in the empirical world. Ashby posited the idea of ultra-stability to explain the changes in the patterns of stability associated with a learning system. This second-order adaptability accounts for the ability of learning organisms (and machines) to change the rules of behavior directing their actions.

These ideas have been integrated in disciplines closely related to the Richardson Tradition. A prominent example is the work of M.D. Mesarovic et al.[24] They have not been picked up by the Richardson Tradition. The work in optimal control theory undertaken by John Gillespie, Zinnes, and others logically leads into the concepts of ultra-stability and second-order adaptation introduced by Ashby and developed by Mesarovic and others. However, this logical progression has not been developed. The relatively high degree of mathematical complexity of optimal control theory is no doubt part of the explanation for this developmental lacuna.

The research reported here takes a different line of attack on these problems. The inordinate complexity of optimal control mathematics for empirical inquiry need not remain a roadblock to progress. We can draw upon alternative and more tractable formalisms capable of supporting ideas about adaptation. The Arms Race Simulation (ARES) model described in subsequent chapters begins this process, taking necessary first steps in the exploitation of these alternative formalisms. Using the first-order production systems formalism, ARES generates complex adaptive behavior in a multi-level model of bureaucratic interaction. More important, however, the insights gleaned in the process suggest that even if the model were enhanced with second-order "meta-rules" for ultra-stable behavior, still we will not have conquered the ghost of mechanism. The problem of intentionality involves degrees of complexity and modes of adaptation that remain unfathomable. The techniques discussed in this work may be capable, eventually, of supporting solutions to this problem. Then again, they may not. It is not yet possible to say, for we don't yet know how to conceptualize the production of subjectivity.

Serious problems with the conceptual adequacy of arms accumulation models will obviously remain unanswered by this study, and therefore the epistemological crisis of the the Richardson Tradition will continue. Possession of an unproblematic and shared exemplar is basic to the identity and legitimacy of a scientific enterprise, and so we see a threat to the scientific legitimacy of the Richardson Tradition. Thus, it is imperative that Richardson's students continue to seek conceptually adequate models of the armaments accumulation phenomenon.

1.4 The Idea of Strategy Is Important

In his often quoted phrase, Richardson described his research project on the mathematical psychology of armaments accumulation as an attempt to represent "what would happen if men did not stop to think." This rather ambiguous aspiration is comprehensible in terms of his general aim of isolating the *mechanical* aspects of war, those "more fated and . . . less freely choosable forms of international behavior."[25] "Not stopping to think" is a recurring metaphor through his entire corpus on the statistical psychology of war. From war moods through reaction coefficients to militant ideologies, Richardson repeatedly refers to the deterministic element of mass affective psychology, rather than a more non-deterministic individual cognitive psychology. All of these notions apparently flow from a deeply rooted belief that anything so repellent to Richardson's Quaker sensibility as is war must certainly involve both a suspension of higher cognitive functions and the abdication of moral responsibility.

Yet there is some vague smell of paradox here. We get a whiff if we interpret Richardson as assuming that, on the occasion of the most perilous of their several occupations, the leaders of a state should lose their grip on that presence of mind that presumably stays their hand in less lethal times. Issues of war and peace arguably are the most vital to confront the state; is Richardson really saying that leaders give this no thought? The scent of paradox dissipates quickly, however, if we view him as having described not what happens when men don't stop to think, but rather *what happens when men think in a certain way.*

That certain way of thinking associated with armaments accumulation is commonly referred to as "strategic thought." Unsurprisingly, strategic thinkers fancy that they think about strategy. But "strategy," defined as the art of devising plans toward a goal, is a strange thing, singularly multifarious. Strategy has a peculiarly interactive character because "what is of supreme importance in war is to attack the enemy's strategy."[26] This is a key notion in the present study: Strategy is by its nature inherently interactive. This fact implies a complex set of relations between political intercourse and force of arms.

Strategy proper has as its object the political will of human beings. The tight web of relations binding strategy to politics is especially crucial in matters of war. "War can never be separated from political intercourse, and if . . . this is done in any way, all the threads of the different relations are, to a certain extent, broken, and we have before us a senseless thing without an object."[27] Under "normal" conditions, force is employed as a means to a political end; under the conditions of armaments accumulation, force becomes an end in itself. The arms race begins as a senseless thing, without a concrete political object and hence divorced from strategy proper.

More than one investigator has argued that "behavioral models [of arms accumulation] would certainly gain more theoretical background and lose some of their mechanistic bias by introducing strategic considerations."[28] Yet, a review of the literature suggests that relatively few analytical studies of armaments accumulation have attempted to examine the role played by strategy in the process.[29] There are several others, but these studies on strategy in arms accumulation typify the whole of this concentration. A fundamental problem with these efforts to import the notion of strategy into armaments accumulation modeling and theory is that they adopt a very simple notion of strategy. Indeed, from the point of view of a strategist, most arms accumulation studies employ reductionist conceptions of strategy.

Strategy is rightly regarded as a complex and in some respects arcane topic. It must be viewed in terms of multiple and interrelated levels of conceptualization. The traditional terminology includes "grand strategy," "military strategy," and "operational strategy." Operational strategy, or simply operations, is usually regarded as referring to the management of the instruments of force in and out of combat, including tactics, logistics, and related matters having to do with the movement of men and machines in physical space. Military strategy is less concrete, involving a higher level of conceptualization that encompasses operations, but goes beyond it to include considerations of the *purpose* of the forces in a politico-military context. Grand strategy is held to be the highest level of conceptualization, reaching in its considerations far beyond mere military force and politico-military relations to include all other factors brought to bear in the creation and execution of the policies of the state in the international environment.

Ironically, arms race theories have mirrored a central dilemma of modern strategic thought. Critics of deterrence theory have argued that this body theory, as it is studied and practiced in contemporary times, is fatally flawed in that it reduces all of strategy to mere operations.[30] This charge is borne of the focus of that theory upon problems of nuclear targeting and the strategic nuclear balance. While it is not my purpose to criticize theories of nuclear deterrence, such criticisms are directly relevant to the study of arms accumulation modeling because arms race theorists have unwittingly adopted much of the conceptual apparatus of deterrence theory in their efforts to understand the nature of the business. In the process, they have focused on the lowest common denominator of military interaction--operations--while slighting the higher levels of conceptualization, which deal with the purposes of the use of force.

It is tragic that the literature on arms accumulation processes often relies upon concepts of strategy as reductionist as, or even more so, than those employed by military bureaucracies. This is reflected in some of the best work in the field. In the innovative and important *Strategy in a Missile War*, Michael D. Intriligator explicitly reduces the idea of strategy to an operations analysis problem for optimizing target selection and rates of fire in nuclear war.[31] While most traditional analysts regard the concept of "grand strategy" as one that necessarily involves the analysis of political goals and the connection between force and will, Dagobert Brito and Intriligator assert that "the problem of *grand strategy* is that of choosing both a rate of fire and a targeting strategy."[32]

Peace researchers recently have been exhorted to consider whether our research methods contribute to war or peace and to wonder under what conditions prophecies (and hypotheses) are self-fulfilling.[33] In the Richardson Tradition these questions have a special urgency. "Normal science" in the Richardson Tradition basically consists of trying to fit a vision of implacable mechanistic conflict interaction onto every example of international interaction having elements of conflict, at best seeking a stable regime of conflict. This focus dishonors the name of peace research, the more so in that it remains generally unrecognized that this harsh vision usually does not fit. When scientific puzzle-solving activity consistently defies solution, the anomaly should give rise to a genuine puzzle.[34] Tragically, the Richardson Tradition has been slow to recognize the existence of anomaly.[35] A more sophisticated conception of strategy is a key element in recognizing and resolving the genuine puzzle of interaction in arms accumulation.

1.5 A Plan of Research Follows from This

The integration of internal and external approaches to arms accumulation theory is implicit in the idea of planning. When defense planners sit down to consider their business, so to speak, they must surely think about internal factors such as production possibility frontiers, the pace of technological innovation, and the spatial distribution of their assets. Likewise, they also certainly must consider external factors such as the disposition, vulnerability, and reactivity of forces against which they may be arrayed. We know that they do consider these things. Diplomatic and military historians and the publishers of the *Congressional Record* make their living writing about it. But just as surely, these internal and external factors are not viewed in isolation, as a set of internal variables counterpoised against a distinct and separate set of external variables. Rather, internal and external factors are considered as they stand in relation to one another, and it is the character of this relation that is crucial to the planning function. For a large modern defense organization, the question of planning becomes what impact a change in the external environment has on one's predefined purposes. Those predefined purposes may have been originally established with respect to an aspect of the environment (e.g., a potential enemy) that is even now responding to the organization's earlier behavior.

Thus, the internal/external dichotomy so popular in academic abstractions largely disappears, or rather is explicitly unified, in the domain of *planning*. Plans reflect the internal goals and external constraints of the organization. Changes in plans reflect the impact of both internal changes in the organization and external changes in the organization's operational environment as they relate to the potential for realization of the goals of the organization. When these internal and external elements are abstracted as mutually exclusive categories in the pursuit of analytical precision, the crucial relationship between the two is excluded *prima facie*. We may therefore propose the scheme in Figure 1.2 as an orienting device for the conduct of this study. Figure 1.2 graphically illustrates the unremarkable contention of this study that armaments and armaments budgets are not directly related in a causal sense or at least that it will be fruitful to proceed as if they were not. Rather, arms and the budgets that pay for them are jointly the product of a set of defense plans. The action of strategic planning in armaments accumulation will be the focus of this study.

It is important to note that the original representation of Richardson's model, as embodied by the system described in his equations, was explicitly designed so that it contains no *intelligence*. Richardson readily admitted that his description "does not mention intelligent aggression planned by a leader as moves in a game of chess," rather focusing upon "the acts of a leader [that] are in part controlled by the great instinctive and traditional tendencies."[36] That is to say, the representation realized in the Richardson reaction model constitutes a wholly mechanical and frankly unrealistic process.

In contrast, it can be argued that Richardson's theory kernel calls for a model with some aspects of intelligence but that he was constrained by the limited techniques available to him to implement a class of representations that fail to capture the richness of his vision. Looking only at the equations, one discerns a determinate, inexorable process. But this distorts the idea. Though Richardson was many things, he was not a pessimist, and he rejected the notion that arms races are somehow inevitable. He held that "nations [are] reluctantly but steadily increasing their armaments" in large part because "they have not yet made a sufficiently strenuous intellectual and moral effort to control the situation."[37] Leaders might learn the better. In the reaction model representation, however, there exists no provision for the adaptation of system structure (diachrony). Learning is limited, at best, to an ability to alter the response set within the context of a pre-given and immutable goal. The Richardson reaction system is locked in.

Richardson does show a theoretical possibility that armaments accumulation could end without fighting, but even then a special "submissiveness" model must be invoked.[38] The basic reaction model--which depicts the entrenching of a policy in a stable bureaucratic regime--does not even admit to the submission solution. But bureaucracies do change, if only slowly. As extensions of human beings, bureaucracies are infused with intentionality. It might be interesting to consider how our models of military interaction would behave if we endowed them with a little of it. Perhaps they might behave a little more like the real thing.

Figure 1.2

Planning as Intervening Variable

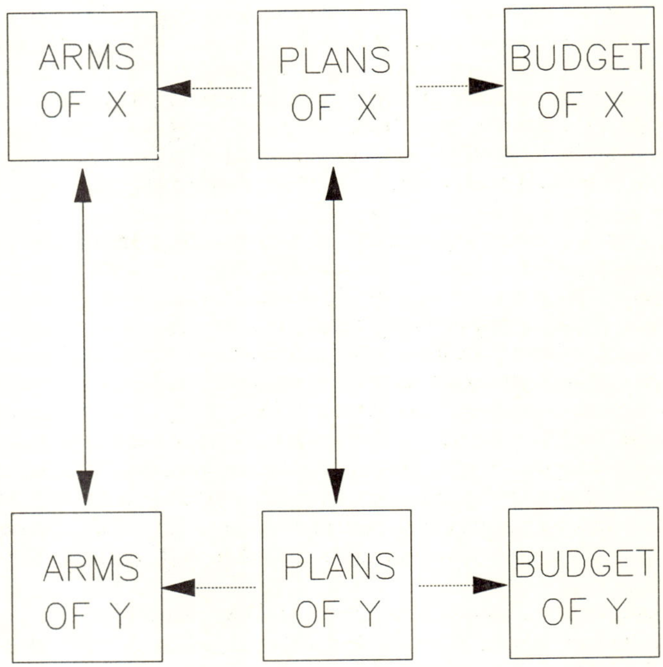

Some attempts have been made to introduce aspects of intelligence into models of foreign policy and defense decision making. These range from the "teaspoon of artificial intelligence" in early studies by Hayward Alker and his collaborators,[39] through the pioneering work of Jamie Carbonell,[40] to more recent attempts to exploit insights and techniques from the field of artificial intelligence in works by a variety of scholars.[41] Work in progress continues by some of these and other investigators. It is within this emergent methodological concentration that the present work should be placed.

1.5.1 Contribution and Limitations

This study is about "changing the system," systems of intentionality. In the process I have sought after the sources of action in armaments accumulation behavior. Believing that the problems of control and purpose lie near the wellsprings of social action, my research focused on those concepts central to the self-defined purposes of the profession of arms: strategy. This focus immediately guides attention to plans, i.e., those rules that provide structure to social behavior. Sets of these rules, implemented with a formalism called the "production system," are at the heart of the ARES model presented here. As will become apparent, the first-order production system approach of ARES leaves some basic problems of arms accumulation modeling untouched.

The production-system approach naturally progresses to a discussion of second-order production systems, which employs a structure known as meta-rules. However, as John Searle points out, techniques for representation of second-order adaptation do not solve the problem of intentionality any better than does the first-order approach.[42] The rules that control the second-order change come, *ex post facto*, from the observations of the analyst. This is true of all adaptive simulation models devised to date. ARES is a prime example. The Ghost lives. In a conceptually adequate model, the model itself would generate these rules for second-order change. Structural change in social reality is not imposed *ex deux machina* by some cosmic modeler. It arises endogenously from physical and social systems themselves. Change is immanent; it comes from within what is. No one has suggested either how this might work or how to model it. I do not do so here.

This study, then, could go either way. The research reported here could be cast as a success--an excursion to regions never visited, specimens carried back--or it could be reported as a failure--reaching too far and never once attaining the ultimate object. There are important elements of both in what follows. I shot for the stars, but only made it to Mars. Yet because no one had visited Mars before, later explorers will better know what to expect out there as a result of this study.

1.6 Summary and Conclusions

We have observed a number of things:

§ *The idea of interaction in armaments accumulation is genuinely puzzling.*

§ *Theories of such interaction are based on the "doctrine of mutuality."*

§ *The mathematics used to represent the idea will influence what can be found.*

§ *The concept of strategy suffers at the hands of theorists and practitioners alike.*

§ *A plan of research follows from this.*

This plan of research is premised upon the problem of interaction in armaments accumulation: Do nations interact in arms races? Chapter One has painted the intellectual history of this problem in very broad strokes. The methods that have been employed in the study of this problem were also given a cursory introduction along with some of their difficulties. One notable theoretical difficulty in this research has been inadequate treatment of the concept of strategy. It was proposed to approach the problem by focusing on planning. This introductory chapter concludes with a prospective summary of the body of this study.

The basis of theories of interaction in armaments accumulation is the little-discussed doctrine of mutuality. The idea of mutuality is often tested as an alternative to the idea of autonomy. Neither autonomy nor mutuality alone is adequate as a conceptualization of armaments accumulation, but synthesized under the concept of strategic planning, they offer interesting analytical possibilities. Strategic planning--the central preoccupation of the ministry of defense--is intentionality in action.

Chapter Two demonstrates that the problem of intentionality has foiled all existing efforts to devise conceptually adequate models of armaments accumulation. Chapter Three suggests that a better understanding of the relationship between theory and method is crucial in any solution to this problem. Chapter Four describes an attempt to come to terms with the problem of intentionality in armaments accumulation interaction by implementing a model using the first-order production system formalism. Chapter Five compares the ARES model presented in the previous chapter to existing structural models and analyzes the inadequacies of the first-order production system formalism for the problem in intentionality in arms accumulation interaction. Chapter Six presents the summary and conclusions. Overall, this work attempts to establish the requirements for an intelligent and conceptually adequate model of strategy and the structure of behavior in armaments accumulation.

Notes

1. James N. Rosenau, "Puzzlement in Foreign Policy," Jerusalem Journal of International Relations 1 (Summer 1976), pp. 1-10.

2. James N. Rosenau, "Thinking Theory Thoroughly," address to the Seminar on International Relations Theory, Jawaharlal Nehru University, New Delhi, May 14-17, 1979. (Typewritten.)

3. Quincy Wright, A Study of War (Chicago: University of Chicago Press, 1942), vol. 2, p. 813.

4. Dina Zinnes, "Three Puzzles in Search of a Researcher," International Studies Quarterly 24 (March 1980), pp. 1-27.

5. Peter Busch, "Mathematical Models of Arms Races," in What Price Vigilance: The Burdens of National Defense, ed. Bruce Russett (New Haven: Yale University Press, 1970), pp. 193-233; Dina Zinnes and John Gillespie, "Analysis of Arms Race Models: USA vs. USSR and NATO vs. WTO," in Modeling and Simulation, eds. W.G. Vogt and M.H. Mickle (Pittsburgh: Instrument Society of America, 1973), vol. 4, pp. 145-148; Hans Rattinger, "Armaments, Detente, and Bureaucracy: The Case of the Arms Race in Europe," Journal of Conflict Resolution 19 (December 1975), pp. 571-595; John Gillespie and Dina Zinnes, "Progressions in Mathematical Models of International Conflict," Synthese 31 (1975), pp. 289-321; W. Ladd Hollist, "Alternative Explanations of Competitive Arms Processes: Tests on Four Pairs of Nations," American Journal of Political Science 21 (May 1977), pp. 313-340; Charles W. Ostrom, Jr., "Evaluating Alternative Foreign Policy Decision-Making Models: An Empirical Test between an Arms Race Model and an Organizational Politics Model," Journal of Conflict Resolution 21 (June 1977), pp. 235-265; Robert Shisko, "Defense Budget Interactions Revisited," Rand Paper P-5882 (Santa Monica, CA: Rand Corp., 1977); Michael Wallace and Judy Wilson, "Non-linear Arms Race Models," Journal of Peace Research 15 (1978), pp. 175-192; Jean-Christian Lambelet, Urs Luterbacher, and Pierre Allan, "Dynamics of Arms Races: Mutual Stimulation vs. Self-Stimulation," Journal of Peace Science 4 (1979), pp. 49-66; Michael Wallace, "The Role of Arms Races in the Escalation of Disputes into War: Some New Evidence," Journal of Conflict Resolution 23 (1979), pp. 3-16; Michael Wallace, "Accounting for Superpower Arms Spending," in Threats, Weapons, and Foreign Policy, eds. Pat McGowan and Charles Kegley (Beverly Hills, CA: Sage, 1980), pp. 259-274; A.F.K. Organski and J. Kugler, The War Ledger (Chicago: University of Chicago Press, 1980); Thomas Cusack and Michael Don Ward, "Military Spending in the United States, Soviet Union, and the People's Republic of China," Journal of Conflict Resolution 25 (September 1981), pp. 429-469; Stephen Majeski and David Jones, "Arms Race Modeling: Causality Analysis and Model Specification," Journal of Conflict Resolution 25 (1981), pp. 259-288; Mike Horn, "Arms Races and the Likelihood of War," paper presented to the annual meeting of the International Studies Association, Atlanta, March 27-31, 1984.

6. Joseph Schumpeter, Imperialism and Social Classes: Two Essays (New York: Signet, 1951).

7. D.A. Davis, M.A.H. Dempster, and Aaron B. Wildavsky, "A Theory of the Budgetary Process," American Political Science Review 60 (1966), pp. 529-547; Dagobert Brito, "A Dynamic Model of an Armaments Race," International Economic Review 13 (1972), pp. 1163-1172; Dagobert Brito and Michael D. Intrilligator, "Some Applications of the Maximum Principle to the Problem of an Armaments Race," in Modeling and Simulation, eds. W.G. Vogt and M.H. Mickle (Pittsburgh: Instrument Society of America, 1974), vol. 4, pp. 140-144; Dagobert Brito and Michael D. Intrilligator, "A Game Theoretical Approach to Bureaucratic Behavior," in Dynamic Optimization and Mathematical Economics, ed. P.T. Liu (New York: Plenum Press, 1980), pp. 223-236; Rattinger, "Armaments, Detente, and Bureaucracy," op. cit.; Marwin Simaan and Jose B. Cruz, Jr., "Nash Equilibrium Strategies for the Problem of Armaments Race and Control," Management Science 22 (September 1975), pp. 96-105; and also their "Formulation of Richardson's Model of Arms Race from a Differential Game Viewpoint," Review of Economic Studies 42 (1975), pp. 67-77; D.L. Wagner, R.T. Perkins, and R. Taagepera, "Fitting the Soviet-U.S. Arms Race Data to the Complete Solution to Richardson's Equations," paper presented to the annual meeting of the Western Regional Peace Science Society, San Diego, February 22, 1974; Stuart Bremer, Simulated Worlds: A Computer Model of National Decision-Making (Princeton, NJ: Princeton University Press, 1977); Ostrom, "Evaluating Alternative Foreign Policy Decision-Making Models," op. cit.; John V. Gillespie, Dina A. Zinnes, and G.S. Tahim, "Deterrence as Second Attack Capability: An Optimal Control Model and Differential Game," in Mathematical Models in International Relations Research, eds. J.V. Gillespie and D.A. Zinnes (New York: Praeger, 1977), pp. 367-385; W. Ladd Hollist and Harold Guetzkow, "Cumulative Research in International Relations: Empirical Analysis and Computer Simulation of Competitive Arms Processes," in Exploring Competitive Arms Processes, ed. W. Ladd Hollist (New York: Marcel Dekker, 1978), pp. 165-197; Dina A. Zinnes, John V. Gillespie, and G.S. Tahim, "The Richardson Arms Race Model," UMAP Unit 308 (Newton, MA: EDC/UMAP, 1978); Majeski and Jones, "Arms Race Modeling," op. cit.; Cusack and Ward, "Military Spending in the United States, Soviet Union, and the People's Republic of China," op. cit.; Michael L. Squires, "Modeling the U.S.-U.S.S.R. Strategic Arms Race" (Ph.D. dissertation, Indiana University, 1982); Stephen J. Majeski, "Mathematical Models of the U.S. Military Expenditure Decision-Making Process" (Ph.D. dissertation, Indiana University, 1981); and also his "Dynamic Properties of the U.S. Military Expenditure Decision-Making Process," Conflict Management and Peace Science 7 (1983), pp. 65-86.

8. Charles W. Ostrom, Jr., "A Reactive Linkage Model of the U.S. Defense Expenditure Policy Making Process," American Political Science Review 72 (1978), pp. 941-957; Rein Taagepera, "Stockpile-Budget and Ratio Interaction Models for Arms Races," Peace Science Society (International) Papers 29 (1980), pp. 67-78; Dagobert Brito and Michael D. Intriligator, "Arms Races: Behavioral and Economic Dimensions," in Missing Elements in Political Inquiry: Logics and Levels of Analysis, eds. Judith Gillespie and Dina Zinnes (Beverly Hills, CA: Sage, 1982), pp. 93-116; Dina A. Zinnes, Barbara J. Hill, David L. Jones, and Stephen J. Majeski, "Modeling Precrisis Interactions," in Missing Elements in Political Inquiry, eds. J.A. Gillespie and D.A. Zinnes (Beverly Hills, CA: Sage, 1982), pp. 117-140; Brian L. Job, "Dynamic Modeling: Introduction and Synthesis," in Missing Elements in Political Inquiry, eds.

J.A. Gillespie and D.A. Zinnes (Beverly Hills, CA: Sage, 1982), pp. 89-92, 141-149; Michael L. Squires, "Modeling the US-USSR Strategic Arms Race," paper presented to the annual meeting of the International Studies Association, Cincinnati, March 25-27, 1982; John Freeman, "Granger Causality and the Time Series Analysis of Political Relationships," American Journal of Political Science 27 (1983), pp. 327-358; Matthew D. McCubbins, "Policy Components of Arms Competition," American Journal of Political Science 27 (1983), pp. 385-406; Michael D. Ward, "The Differential Paths to Parity: A Study of the Contemporary Arms Race," American Political Science Review 78 (1984), pp. 297-317; and Stephen J. Majeski, "Expectations and Arms Races," paper presented to the annual meeting of the International Studies Association, Atlanta, March 27-31, 1984.

9. Lewis Fry Richardson, Arms and Insecurity: A Mathematical Study of the Causes and Origins of War (Pittsburgh: Boxwood Press, 1960), p. 62.

10. Klause Jurgen Gantzel, "Armament Dynamics in the East-West Conflict: An Arms Race?" Peace Science Society (International) Papers 20 (1973), p. 1.

11. Johan Jorgen Holst, "What Is Really Going On?" Foreign Affairs 19 (Summer 1975), pp. 155-163.

12. T.A. Brown, "What Is an Arms Race?" (Santa Monica, CA: Rand Corp., Southern California Arms Control and Foreign Policy Seminars, 1973); Albert Wohlstetter, "Is There a Strategic Arms Race?" Foreign Policy 15 (Summer 1974), pp. 3-20; and his follow-on piece, "Rivals, but No 'Race,'" Foreign Affairs 16 (Fall 1984), pp. 48-81; also interesting here is Wohlstetter's "Racing Forward or Ambling Back," in Defending America, eds. James R. Schlesinger et al. (New York: Basic Books, 1977), pp. 110-168.

13. Lewis Fry Richardson, "Threats and Security," in Psychological Factors in Peace and War, ed. T.H. Pear (London: Hutchinson and Co., 1950), p. 233.

14. Together, Equations [1.1] and [1.2] comprise the basic action-reaction model of arms accumulation, where x and y represent the arms of two nations, dx/dt and dy/dt represent the rate of change in their level of armaments, and a, b, k, l, g, and h are constants representing their respective reactivity to arms of the other (k and l), their economic burdens of arms spending (a and b), and their underlying grievance and/or hostility (g and h).

15. Squires, "Modeling the US-USSR Strategic Arms Race," op. cit., p. 212.

16. Majeski, "Expectations and Arms Races," op. cit.

17. G.W.F. Hegel, The Philosophy of Right, trans. T.M. Knox (Chicago: Encyclopaedia Britannica, 1952), p. 323.

18. Max Weber, Economy and Society, trans. A.M. Henderson and T. Parsons (Glencoe, IL: Free Press, 1957).

19. Thomas Schelling, Arms and Influence (New Haven: Yale University Press, 1966).

20. M.D. Salomon, "New Concepts for Strategic Parity," Survival 19 (1977), pp. 255-262.

21. Squires, "Modeling the US-USSR Strategic Arms Race," op. cit., p. 225.

22. Charles W. Ostrom, Jr., "The Politics of the U.S. Defense Expenditure Policymaking Process: An Empirical Test between an Arms Race Model and an Organizational Politics Model" (Ph.D. dissertation, Indiana University, 1975); Philip A. Schrodt, "The Effects of a Richardson Arms Race on the Distribution of Power in an International System" (Ph.D. dissertation, Indiana University, 1976); Majeski, "Mathematical Models of the U.S. Military Expenditure Decision-Making Process," op. cit., 1981; Squires, "Modeling the US-USSR Strategic Arms Race," op. cit.

23. W. Ross Ashby, An Introduction to Cybernetics (New York: John Wiley and Sons, 1956); and his Design for a Brain: The Origin of Adaptive Behavior (New York: John Wiley and Sons, 1960).

24. M.D. Mesarovic et al., Theory of Hierarchical, Multilevel Systems (New York: Academic Press, 1971).

25. Richardson, Arms and Insecurity, op. cit., p. xxxv; cf. David Wilkenson, Deadly Quarrels: Lewis F. Richardson and the Statistical Study of War (Berkeley: University of California Press, 1980), p. 7.

26. Sun Tzu, The Art of War, trans. D. Griffith (London: Oxford University Press, 1963).

27. Karl Von Clausewitz, On War, trans. Anatol Rapoport (Princeton, NJ: Princeton University Press, 1968), p. 402.

28. Urs Luterbacher, "Towards a Convergence of Behavioral and Strategic Conceptions of the Arms Race: The Case of the American and Soviet ICBM Build-up," Peace Science Society (International) Papers 26 (1976), p. 5.

29. Brito and Intriligator, "Some Applications of the Maximum Principle to the Problem of an Armaments Race," op. cit.; also their "Uncertainty and the Armaments Race," Annals of Economic and Social Measurement 3 (1974), pp. 279-292; and especially their "Arms Races: Behavioral and Economic Dimensions," op. cit.; Michael D. Intriligator, "Some Simple Models of Arms Races," General Systems 9 (1964), pp. 143-147; his "Strategy in a Missile War," Paper No. 10 (Los Angeles: UCLA Security Studies Project, 1967); also his "The Debate over Missile Strategy: Targets and Rates of Fire," Orbis 11 (1968), pp. 1138-1159; and his "Applications of Optimal Control Theory in Economics," Synthese 31 (1975), pp. 271-288; Michael D. Intriligator and Dagobert Brito, "Strategy, Arms Races, and Arms Control," in Mathematical Systems in International Relations, eds. J.V. Gillespie and D.A. Zinnes (New York: Praeger, 1977), pp. 173-189; Gantzel, "Armaments Dynamics in the East-West Conflict: An Arms Race?" op. cit.; John V. Gillespie et al., "Deterrence as Second Attack Capability: An Optimal Control Model and Differential Game," op. cit.; Robert H. Kupperman and Harvey A. Smith, "Strategies of Mutual Deterrence," Science 176 (April 1972), pp. 18-23; and their "Deterrent Stability and Strategic Warfare," in Mathematical Systems in International Relations, eds. J.V. Gillespie and D.A. Zinnes (New York: Praeger, 1977), pp. 139-166; Luterbacher, "Towards a Convergence of the Behavioral and Strategic Conceptions of the Arms Race," op. cit.; and Martin C. McGuire, Secrecy and the Arms Race: A Theory of the Accumulation of Strategic Weapons and How Secrecy Affects It (Cambridge, MA: Harvard University Press, 1965).

30. See, for example, Salomon, "New Concepts for Strategic Parity," op. cit.

31. Intriligator, "Some Simple Models of Arms Race," op. cit.

32. Michael D. Intriligator and Dagobert Brito, "Heuristic Decision Rules, the Dynamics of the Arms Race, and War Initiation," paper presented to the Joint Seminar on Dynamic Models of International Conflict, Boulder, CO, October 31 to November 3, 1984, p. 6; emphasis in original.

33. William Eckhardt, "Peace Research Agenda for the 1980's," International Studies Newsletter 11 (June/July 1984), p. 15.

34. Rosenau, "Thinking Theory Thoroughly," op. cit.; Zinnes, "Three Puzzles in Search of a Researcher," op. cit.

35. Richard K. Ashley, "Noticing Pre-paradigmatic Progress," in In Search of Global Patterns, ed. J.N. Rosenau (New York: Free Press, 1976), pp. 150-158.

36. Richardson, Arms and Insecurity, op. cit., p. 227.

37. Ibid., p. 12.

38. Lewis Fry Richardson, "Could an Arms Race End without Fighting?" Nature 167 (September 29, 1951), p. 567.

39. Hayward R. Alker, Jr., and C. Christensen, "From Causal Modeling to Artificial Intelligence: The Evolution of a UN Peace-Making Simulation," in Experimentation and Simulation in Political Science, eds. J.A. LaPonce and Paul Smoker (Toronto: Toronto University Press, 1972); James P. Bennett and Hayward R. Alker, Jr., "When Collective Security Policies Bred Collective Insecurity: The War of the Pacific in a World Politics Simulation," in Problems of World Modeling, eds. Karl W. Deutsch, Bruno Fritsch, Helio Jaguaribe, and Andrei Markovits (Cambridge, MA: Ballinger Publishing Co., 1977), pp. 215-302.

40. Jamie G. Carbonell, "POLITICS: Automated Ideological Reasoning," Cognitive Science 2 (1978), pp. 27-51; also his "The Counterplanning Process: A Model of Decision-Making in Adverse Situations" (Pittsburgh: Carnegie-Mellon University, Department of Computer Science, February 1979); and also his "POLITICS," in Inside Computer Understanding, eds. R. Schank and C. Riesbeck (Hillsdale, NJ: Lawrence Erlbaum, 1981), pp. 259-307.

41. Recently this literature has begun to explode in breadth and depth, including Paul A. Anderson and Stuart J. Thorson, "Artificial Intelligence Based Simulation of Foreign Policy Decision-Making," Behavioral Science 27 (April 1982), pp. 176-193; Sanjay Banerjee, "Technical Knowledge in a Political Economy Context: An Artificial Intelligence Based Model," paper presented to the annual meeting of the International Studies Association, Anaheim, CA, March 25-29, 1986; Scott Bennett and Philip A. Schrodt, "Linear Event Patterns in WEIS Data," paper presented to the annual meeting of the American Political Science Association, Chicago, September 3-6, 1987; Davis B. Bobrow, Donald A. Sylvan, and Brian Ripley, "Japanese Supply Security: A Computational Model," paper presented to the annual meeting of the International Studies Association, Anaheim, CA, March 25-29, 1986; Robert Boydston, "Research on Non-violent Strategies: Artificial Intelligence as a Methodological Tool," paper presented to the annual meeting of the International Studies Association, Washington, DC, March 5-9, 1985; Gavan Duffy and John Mallery, "RELATUS: An Artificial Intelligence Tool for Natural Language Modeling," paper presented to the annual meeting of the International Studies Association, Anaheim, CA, March 25-29, 1986; Margee M. Ensign, "Artificial Intelligence-Based Simulations in International

Relations," paper presented to the annual meeting of the International Studies Association, Anaheim, CA, March 26-29, 1986; Craig Etcheson, "Expert Systems for Country Risk Analysis," paper presented to the annual meeting of the International Studies Association, Atlanta, March 27 to April 1, 1984; Valerie M. Hudson, "Some Issues in the Application of Artificial Intelligence Methodology to the Study of International Relations," paper presented to the annual meeting of the International Studies Association, Washington, DC, March 5-9, 1985; Brian Job and Douglas Johnson, "A Formal Decision-Making Model of U.S. Intervention: The U.S. and the Dominican Republic," paper presented to the annual meeting of the International Studies Association, Anaheim, CA, March 25-29, 1986; Brian Job, Douglas Johnson, and Eric Selbin, "United States Intervention in Central America: A Model and Application to the Dominican Republic, 1959-1965," paper presented to the annual meeting of the American Political Science Association, Chicago, September 3-6, 1987; Ruth Lane, "Artificial Intelligence and the Political Construction of Reality: The Case of James E. Carter," paper presented to the annual meeting of the American Political Science Association, Washington, DC, August 28-31, 1986; Stephen Majeski and David Sylvan, "Decision Rules for War Involvement: Vietnam Revisited," paper presented to the annual meeting of the International Studies Association, Washington, DC, March 5-9, 1985; and also Majeski and Sylvan's "A Computational Model of the U.S. Foreign Policy Recommendation Process: Vietnam and Laos in 1961," paper presented to the annual meeting of the American Political Science Association, Chicago, September 3-6, 1987; John C. Mallery, Gavin Duffy, and Roger Hurwitz, "Computing Strategic Language: Natural Language Models of Belief and Intention," paper presented to the annual meeting of the International Studies Association, Washington, DC, April 14-18, 1987; John C. Mallery and Roger Hurwitz, "Analogy and Precedent in Strategic Decision-Making: A Computational Approach," paper presented to the annual meeting of the American Political Science Association, Chicago, September 3-6, 1987; Richard Molnar, "Modelling Cognitive Constraints in Decision-Making: An Application to U.S. Foreign Policy," paper presented to the annual meeting of the International Studies Association, Washington, DC, April 14-18, 1987; David Sylvan and Stephen Majeski, "Multiple Actor Advisory Simulations in Foreign Policy," paper presented to the annual meeting of the International Studies Association, Anaheim, CA, March 25-29, 1986; Donald A. Sylvan and Steve Chan, eds., Foreign Policy Decision Making: Perception, Cognition, and Artificial Intelligence (New York: Praeger, 1984); James N. Rosenau, "The Microelectronic Revolution and the Conduct of Foreign Policy," paper presented to the annual meeting of the American Political Science Association, Washington, DC, August 30 to September 2, 1984; Howard Tamashiro, "Algorithms, Heuristics, and Artificial Intelligence Modelling of Strategic Statecraft," in Foreign Policy Decision Making: Perception, Cognition, and Artificial Intelligence, eds. Donald A. Sylvan and Steve Chan (New York: Praeger, 1984), pp. 197-226; Akihiko Tanaka, "China, China Watching, and CHINA_WATCHER," in Foreign Policy Decision Making: Perception, Cognition, and Artificial Intelligence, eds. Donald A. Sylvan and Steve Chan (New York: Praeger, 1984), pp. 310-344; finally, see the bibliography at the end of this volume for an extensive collection of papers by James Bennett, Dwain Mefford, Phil Schrodt, Stuart Thorson, and their collaborators concerning AI applications.

42. John R. Searle, "Minds, Brains, and Programs," in Mind Design: Philosophy, Psychology, Artificial Intelligence, ed. John Haugeland (Cambridge, MA: MIT Press, 1981), p. 303.

2 THEORIES OF INTERACTION IN ARMS ACCUMULATION

The problem of interaction in armaments accumulation has generated a continuous stream of scientific puzzles for students of the Richardson Tradition across most of the twentieth century. How have these scientific pioneers conceptualized the idea of interaction in armaments accumulation? What kinds of explanations have they adduced in attempts to satisfy their curiosity about this phenomenon? How have students of the Richardson Tradition attempted to incorporate ideas about strategy and bureaucracy into their models? This chapter explores these questions by means of a review of the literature on Richardson models of arms races.

The progress of the Richardson Tradition over the past seventy years can be seen as a groping search to conceptualize clearly the empirical reality of armaments accumulation processes, a search that continues to expand into new spaces of knowledge. Pioneers of the Richardson Tradition extended scientific attention to arenas never before subjected to such analytical scrutiny. There is much uncharted territory ahead awaiting explorers of the Richardson Tradition. But for now we shall go back to the very beginning of the search.

2.1 If They Did Not Stop to Think . . .

The guns of August 1914 heralded the first great global conflict in the Age of Total War, and as the ways of war changed so did our ways of thinking about war change. Though all Europe seemed to rush gladly headlong into the coming madness, in fact there were many who moved to resist unleashing the dogs of war. One such was the philosopher Bertrand Russell. Joining with Norman Angell and others, Russell formed the Union of Democratic Control to agitate against the war policies of the British government. Among the polemical tracts published by the Union was one called *War, the Offspring of Fear*. He wrote,

> In every war it appears to each side that it is fighting to resist unprovoked aggression. . . . Let us try for a moment to forget praise and blame . . . and to view the whole tragic irony from the standpoint of impartial compassion and understanding . . . each nation believes that it is defending a sacred cause . . . the one motive that makes populations acquiesce is Fear. . . . If civilization is to continue, Europe must find a cure for this universal reign of fear with its consequence of mutual butchery.[1]

Scholars have long singled out fear as a central element in the coming of war; more than 2,400 years ago Thucydides argued that the Spartans' fear of Athens made the Peloponnesian War "inevitable." But Russell's idea--that the cause of war is mutual fear--was revolutionary, and it had a revolutionary impact on the study of war. That simple propaganda pamphlet sparked what has come to be known as the "Richardson Tradition of arms race analysis." This chapter is devoted to a review of the scientific literature produced by that tradition, with particular attention to attempts to develop theories of armaments accumulation.

2.1.1 The Doctrine of Mutuality

A distinguished young British scientist, Lewis Fry Richardson read Russell's anti-war pamphlet with keen interest. The notion that the motives of each side in a conflict were essentially the same--each acting in fear of the other--and that this fear caused each to raise armaments for defense against the other, formed the foundation of Richardson's analysis of the genesis of war. Richardson went beyond Russell's idea, abstracting what he saw as the essence of the relationship: "When we analyze arms races it is, however, unnecessary to mention fear, or any other emotion.... The valuable part of Russell's doctrine was not his emphasis on fear, but his emphasis on mutual stimulation."[2]

Before the end of the First World War, Richardson had penned the seminal contribution to the tradition, which would come to bear his name. His *Mathematical Psychology of War*, published in 1919, presented the mathematical formulae embodying Russell's "Doctrine of Mutuality"--a mutual stimulation process leading nations to war. Follow-on studies by disciples of Richardson today number in the hundreds, but that first contribution lay dormant and largely unnoticed for some forty years.

It was only in the late 1950s that Richardson's work began to attract a lively interest. Popularized by a group of scholars that included Gregory Bateson, Quincy Wright, Nicholas Rashevsky, Anatol Rapoport, Kenneth Boulding, and Ernesto Trucco,[3] the Richardson Tradition is now recognized as one of the most carefully developed and exhaustively studied areas in the so-called scientific study of international relations.[4] In the twenty-five years since the rediscovery of Richardson's work, a substantial quantity of research has been done on the original set of problems presented by Richardson and on subsequent problems arising in the tradition. The studies of the Richardson Tradition--which number certainly more than six hundred and probably fewer than one thousand--have been conducted by scores of international relations scholars and other social scientists. The present chapter is an analysis of this voluminous literature, based on a sample in excess of 600 scholarly papers and books, encompassing the work of almost 250 scholars. Approximately 97% of this literature appeared within the past twenty-five years. Figure 2.1 displays the distribution of this literature in time. Of this sample, some 75% was collected and analyzed in support of this study (see Appendix A).

The question of defining the boundaries of this literature is a vexing one. There have been many developments and variations on the original work by Richardson, some so different from Richardson's work as to throw into doubt the assertion of their lineage. I began by defining the Richardson literature as including any study that cited Richardson. This quickly proved insufficient.

Figure 2.1 The Richardson Literature

Frequency, 1914–1987 (N=606)

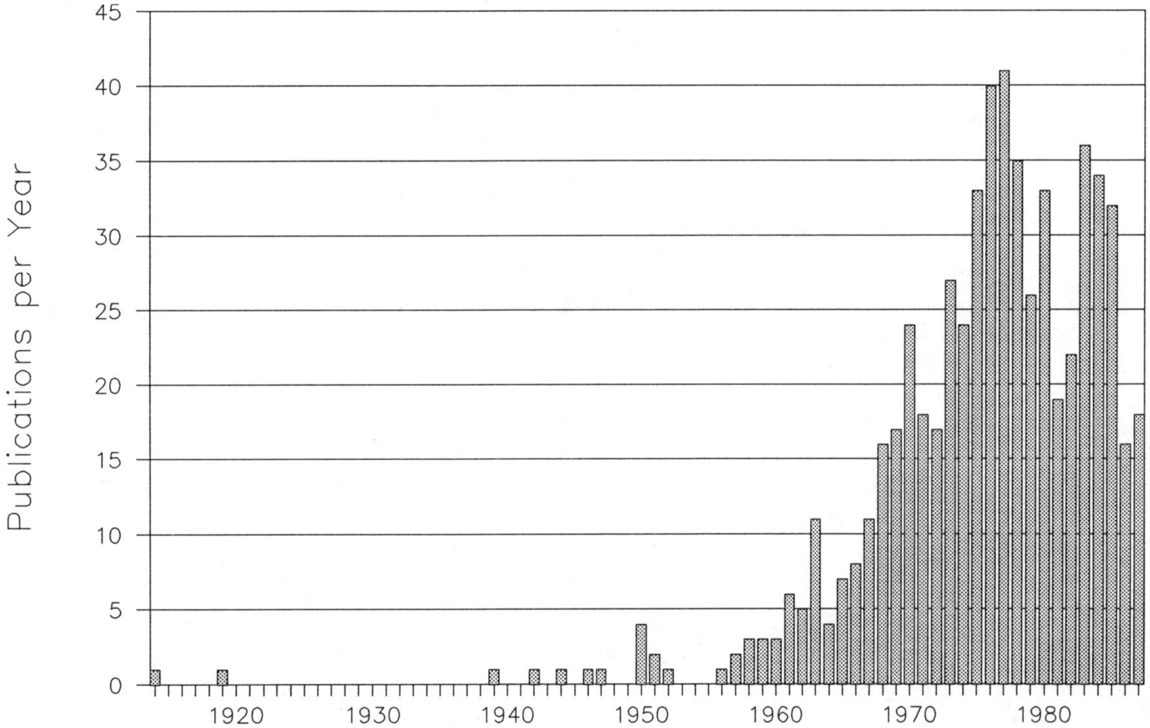

Many papers that clearly belong in the tradition do not cite the original contributions. A few authors have extraneous citations to Richardson while writing on topics unconnected with the study of armaments accumulation. My working definition of the Richardson Tradition literature then collapsed into a series of frankly unsystematic judgment calls, as I attempted to gather all methodological, theoretical, and empirical studies concerned with aspects of armaments accumulation. This procedure resulted in some hard calls. For example, John Nash's article on the bargaining problem is included because of its influence on Brito, Intriligator, Marwin Simaan, Jose Cruz, Gillespie, Zinnes, and others.[5] In contrast, the literature on Lanchester-type combat models (with a few salutatory exceptions) was not included because it has a distinct origin and so distinctly differs in normative orientation from the Richardson Tradition as to qualify as a rival tradition. Overall, however, I sought to err on the side of inclusivity, with the result that some of the papers included in the listing of the Richardson Tradition are sure to provoke debate, in some cases perhaps even by the authors in question. Where to cut off the boundaries of the literature is made more difficult by the fact that in important ways Richardson is the forerunner of all who engage in quantitative approaches to the scientific study of international politics.

2.1.2 The Origins of a Tradition

Born on October 11, 1881, to petty bourgeois parents at Newcastle-Upon-Tyne in the County of Durham, England, Lewis Fry Richardson was raised a Quaker. His early education at the Newcastle Preparatory School and Bootham School "taught him the pleasures of Euclid," and gave him "glimpses of the marvels of science."[6] His Quaker masters inculcated in Richardson the strong conviction that "science ought to be subordinate to morals."[7] In 1900 he entered King's College, Cambridge, where he studied physics under the discoverer of the electron, Professor J.J. Thomson. In 1903 Richardson earned first-class honors in the Natural Science Tripos.[8] While at Cambridge he decided, Richardson later wrote, "I would . . . spend the first half of my life under the strict discipline of physics, and afterwards apply that training to researches on living things. I kept this program a secret."[9] And he did precisely that, which was probably for the best, because the notion of applying the methods of the physical sciences to social problems was virtually unheard of. His early work in this area was ridiculed by many of his scientific colleagues.

In the decade after completing his formal education, Richardson held a variety of posts in academia, government, and private industry, until in 1913 he was appointed as superintendent of the Eskdalemuir Observatory. He resigned that prestigious post three years later in order to volunteer for service in the Friends Ambulance Unit attached to the 16th French Infantry Division. The next three years he spent near the front lines of the war in France. It was the key period of his life, for something about the experience of being close to combat excited Richardson's fertile brain to its utmost. During his non-violent service in the war, his most enduring scientific contributions crystalized and developed. Interestingly, he employed one and the same insight to make his contributions both to meteorology and to peace science. His 1922 classic, *Weather Prediction by Numerical Process*, was the first to elucidate the thermodynamics of the atmosphere, more than a generation prior to development of the observational and computational techniques necessary to corroborate his theory. E. Gold, in the obituary prepared for the Royal Society of Fellows upon Richardson's death, succinctly summarized the theory kernal informing Richardson's corpus on weather prediction: "Turbulence will increase if the energy gained exceeds the energy lost."[10] It was a remarkably simple insight, one precisely analogous to the idea underlying his models of arms races and war.

The philosophers of the Enlightenment who envisioned the Newtonian Universe as a vast mechanistic clockwork had an influence that still dominates the work of contemporary social and physical sciences. The scientific contributions of Tycho Brahe, Johannes Kepler, Galileo Galilei, and Isaac Newton to the study of celestial mechanics, both in the beauty of their method and the utility of their results, created a paradigmatic model of ideal scientific inquiry that directly influenced Richardson's work. "Newton's elementary theory, in his *Principia*, Book 3, Proposition 4," Richardson was at pains to point out, "represented the moon as moving uniformly in a circle, neglected the sun, and was obviously inaccurate. Newton himself made elaborations; and several other astronomers have improved the agreement between theory and observation."[11] So it would be with Richardson's elementary model of arms races.

Richardson's training in natural science at Cambridge had been distinctly positive in character, steeped in the prevailing norms of methodological monism, mathematical idealism, and a search for law-like behavior under the hypothetico-deductive method.[12] Galileo slew Aristotle's impetus theory of motion by abstracting motion as a particle in Euclidean geometric space. True to his positivist epistemology, Richardson mechanically applied the Galilean method lock, stock, and barrel in an attempt to slay the theory of *Realpolitik.* But in the scientific revolution that overthrew Aristotilian physics and Ptolemic astronomy, Brahe gathered the observational data,[13] Kepler deduced the laws of planetary motion,[14] Galileo gathered all the threads into a funeral shroud for Aristotle,[15] and only then could Newton synthesize the whole into a convincing, unified, and triumphant scientific paradigm.[16] Richardson tried to do it all, all by himself.

Thinking in a highly abstract fashion, Richardson imagined that arms races in social systems might be analogous to storms in atmospheric systems and amenable to treatment with methods identical to those he used in the numerical prediction of weather. This analogy enabled him to posit effects of the doctrine of mutuality that could be tested. The doctrine of mutuality implies a synergic effect on the international system, so Richardson posited an entity called the "warpeace particle." This particle exists in a space of particles, the "international plane" or "warpeace space," which represents all possible states or conditions of the system (i.e., a state space). The synergism is the product of the interaction of two countries mutually responding to the attitude of threatening or cooperation of the other. In the case of an arms race, the mutual interaction adds energy to the system, accelerating the momentum of the particle in the warpeace space.

The warpeace space can be visualized as a two-dimensional graph with the attitude of threatening or cooperation of country A plotted along the x-axis, and that of country B along the y-axis, as in Figure 2.2. Positive values represent hostility, negative values

Figure 2.2 The Warpeace Space

Graph of dy/dt = lx, for l = .4

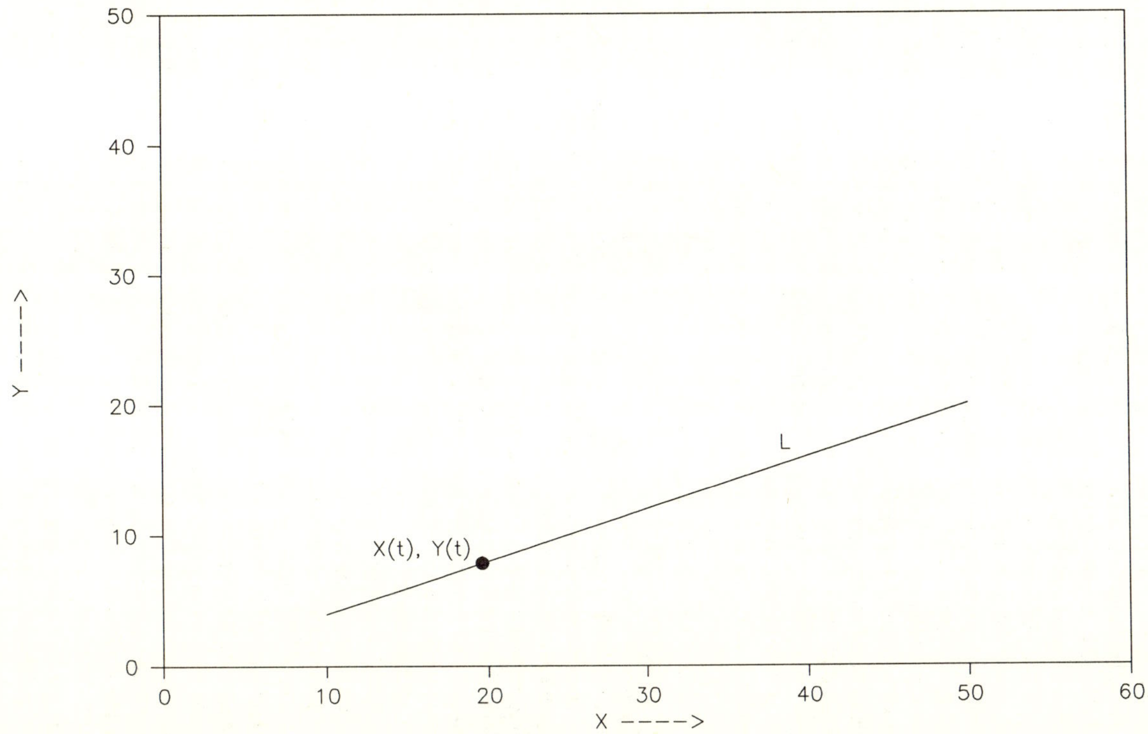

represent cooperation. The location of the warpeace particle in the state space at any given point in time is a function of the mutual attitudes of countries A and B at that time, as in point (x^t, y^t) on Figure 2.2. A plot of these mutual attitudes over time will trace the dynamic motion of the particle in warpeace space, as does line L on Figure 2.2. The principal source of scientific puzzles for the Richardson Tradition has revolved about finding a control structure that describes the dynamic motion of the warpeace particle in international plane.

"What are the relations between the various causes of war? There are plenty of types of relations to choose from, namely all the formulae in the experimental sciences and presumably others not yet discovered."[17] Richardson's initial conjecture about the motion of the warpeace particle came from a purely formal interpretation of the doctrine of mutuality. He envisioned this mutuality as the simple product of the attitude of the other (say, x and y, respectively) multiplied by a factor that represented the degree of reactivity of the one to the other (say, k and l), called the "defense coefficient." Thus, he could represent the rate of change in the attitude of country A, dx/dt (using the Leibnitzian notation), and similarly for country B, with the following equations:

$$dx/dt = ky \qquad [2.1]$$

$$dy/dt = lx \qquad [2.2]$$

These two equations represent Richardson's elementary theory of arms races, the basic mechanistic reaction model of armaments accumulation devised in 1919.[18] Since this initial conjecture, the history of the Richardson Tradition of armaments accumulation studies has been the story of attempts to collect evidence about the doctrine of mutuality by developing elaborations on this elementary model, that is, by attempting to specify the mathematical control structure that describes the path of the warpeace particle in the international plane over time.

Richardson himself experimented with several mathematical structures, including analytical geometry and probability theory, but principally he worked with differential equations. His students have replicated his work and expanded it. Various formal mathematical structures have been suggested and some have been tested in a variety of empirical domains. In a review of nearly two hundred works, Claudio Cioffi-Revillia catalogs a large number of mathematical structures that have been applied to this problem.[19] These include combinatorics, ordinary sets, difference equations, stochastic processes, utility structures, games (discrete deterministic, sequential, and stochastic, continuous, differential, and embedded), graphs, digraphs, networks, linear algebra, equalities and inequalities, analytic geometry, ordinary and partial differential equations plus integro-differentials and other hybrids, topology, continuous systems, control theory, static and dynamic optimization, and programming.

As Cioffi-Revillia notes, however, most of the work has been concentrated on a few particular mathematical structures, with discrete and continuous systems of difference and differential equations accounting for nearly half of the literature he reviews. The dominance of differential and difference equation-type models in the Richardson Tradition is a natural reflection of the fact that this was Richardson's original model-building language of choice.[20] In fact, of the literature sample reviewed in this chapter, more than 51% of the mathematical treatments involved difference or differential equation models (see Table 2.1). Thus, it is natural that the literature review should begin with this approach.

2.1.3 Linear Theory of Two Nations

Richardson's elementary model (represented in Equations 2.1 and 2.2) of arms race mutual stimulation was obviously inaccurate, just as was Newton's elementary model of planetary motion. The unmediated positive feedback system formed by the two equations is explosively unstable, such that even very small defense coefficients (k and l) lead to an extreme escalation process that we do not tend to see in the empirical world. It seemed necessary to add something to the elementary model in order to dampen this explosive relation and bring the model more into line with reality.

Richardson solved this problem by imagining that with increasing expenditures on armaments, soon the effect of diminishing marginal returns on armaments investment would come into play in the system, slowing or damping the interaction process. He hypothesized that this damping effect was directly proportional to the amount invested in arms: the quantity (-ax) defines the so-called fatigue or economic constraint variable for nation X, such that x represents the military expenditures of nation X and -a represents the proportional effect of diminishing marginal returns on military investment. The arms race model then is given as:

$$dx/dt = ky - ax \qquad [2.3]$$

$$dy/dt = lx - by \qquad [2.4]$$

These equations define a pair of slopes in the international space. When one adds to each equation an additional term, the intercept term, one has described a particular pair of lines in the international space. The intercept term was interpreted by Richardson as giving the amount of "hostility" or "grievance" in the system, i.e., the amount of armaments a nation would possess even if there were no effect of mutual stimulation between the two nations. With the addition of the hostility constant, Richardson completed the specification of what has come to be known as the "basic action-reaction model" of armaments interaction. We will refer to this model as the "Basic Model":

$$dx/dt = ky - ax + g \qquad [2.5]$$

$$dy/dt = lx - by + h \qquad [2.6]$$

The Basic Model has been the subject of the greatest bulk of interest in the Richardson Tradition, but Richardson posited three additional derivative models that also have been of considerable interest to his students. The first of these is known as the "Combined Model," formed simply by adding the two equations of the Basic Model together. This yields:

$$d(x+y)/dt = (k - a)(x + y) + g + h \qquad [2.7]$$

Table 2.1 Mathematical Approaches in the Richardson Tradition

Structure or Approach	Frequency
Difference/Differential Equations	154
Probability Models	43
Simulation Approaches	38
Game Theory	27
Optimization	22
Analytical Geometry	20
Equalities and Inequalities	7
Bi-variate Regression	7
AI Approaches	5
Events Data	3
Factor Analysis	2
Topology	2
Logit Analysis	2

Source: RICHLIT DBMS runs on April 14 and 18, 1985. Refer to Appendix A for additional information.

Note: Of the 369 works reviewed, 301 (81.6%) involved mathematical treatment; total sum above is greater than 301 because some authors employ multiple approaches.

Exactly why Richardson chose to formulate the Combined Model remains subject to some considerable debate.[21] There are several questions about the interpretation of this model that remain to be answered in a satisfactory fashion.

Two further elaborations of the Basic Model are found in the Submission Model and the Rivalry Model. The Submission Model is given as:

$$dx/dy = ky\{1 - q(y - x)\} - ax + g \qquad [2.8]$$

$$dy/dt = lx\{1 - p(x - y)\} - by + h \qquad [2.9]$$

Here the submission coefficients q and p reference the degree to which nations X and Y are submissive, that is, the extent to which the nation responds to arms increases of the opposite number with decreases in their own rate of armament accumulation. Richardson offered the Submission Model in an attempt to illustrate how an armaments accumulation interaction might be mitigated prior to the outbreak of fighting. If the submission coefficients are equal to zero, then the Submission Model reduces to the Basic Model. The Submission Model has been treated by Richardson, Paul Smoker, and Zinnes and Gillespie.[22]

The final model offered by Richardson was the Rivalry Model, in which we suppose "that what moves a government to arm is not the absolute magnitude of other nations' armaments but the difference between its own and theirs."[23] In this model the first term on the right-hand side of the Basic Model is replaced with a term defining the difference in the expenditures of two nations:

$$dx/dt = k_1(y - x) - a_1x + g \qquad [2.10]$$

$$dy/dt = l_1(x - y) - b_1y + h \qquad [2.11]$$

Investigators dealing with the Rivalry Model include Richardson, William Caspary, Zinnes and Gillespie, Partha Chatterjee, Hans Rattinger, and Farid Abolfathi.[24]

Richardson's five principal arms race models--the Elementary Model, the Basic Model, the Combined Model, the Submission Model, and the Rivalry Model--have had the benefit of extensive secondary investigation. By far and away, however, the Basic Model has received the vast bulk of analytic attention from the students of the Richardson Tradition. In its unadulterated form, the Basic Model has been considered from a theoretical point of view by several investigators.[25] Beyond purely theoretical considerations, numerous investigators have applied the Basic Model to empirical investigation of specific conflict situations. These include World War I,[26] World War II,[27] the Cold War, i.e., U.S.-Soviet interaction,[28] NATO versus the Warsaw Pact,[29] Israeli-Arab conflict,[30] the United States and Vietnam,[31] the India-Pakistan conflicts,[32] the Sino-Indian conflict,[33] the major powers of the twentieth century,[34] and the United States alone.[35]

Working with modified formulations of the Basic Model, other investigators have attempted to push the explanatory value of the model into uncharted domains. Theoretical work on modified versions of the Basic Model has been undertaken by Richard Brody, Marlin Patchen, Brito, Simaan and Cruz, Gillespie and Zinnes, Dieter Ruloff, Squires, Dragoslav Siljak, and Richard Stoll.[36] As with the pure Basic Model, empirical work on modified versions of the Basic formulation has been undertaken for a variety of conflict situations. These include World War I,[37] World War II,[38] NATO versus the Warsaw Pact,[39] Israeli-Arab conflicts,[40] U.S.-Soviet interaction,[41] the India-Pakistan Wars,[42] British-Irish conflict,[43] and the Iraq-Iran War.[44] With a version of the Basic Model altered to accommodate the three-nation case, the Sino-Soviet-American triangular relationship has been examined by J. Despres and P. Drhymes, Robert Shisko, John Hunter, Alex Hybel and Dale Robertson, Richard Ashley, and Thomas Cusack and Michael Ward.[45]

Altogether, this represents a substantial amount of analytical activity directed towards the investigation of Richardson's Basic Model of arms races. In these several studies, almost every imaginable conclusion is reached at one time or another. A recurrent theme does, however, emerge. In studies finding both that the Basic Model seemed to fit a particular situation as well as studies finding that the Basic Model did not seem to fit at all, analysts frequently ended up by foundering in their attempts to interpret the "defense coefficient." Beyond the tautological mathematical laws governing the behavior of the system of equations, what are the empirical referents of the abstract representation of motion of the warpeace particle in the international space? The absence of a satisfactory answer to this question motivated the Richardson Tradition of armaments analysis to a second major phase of investigations.

2.2 Leaders Lead People Where They Are Willing to Be Led

At the risk of oversimplifying a vast and complex literature, one general conclusion that emerges from a careful reading of the theoretical and empirical treatments of Richardson's Basic Model in its original and modified forms is that it does not provide a satisfying explanation of armaments accumulation processes. This result has evoked a variety of responses from students of the Richardson Tradition. Some have attempted to respecify the Basic Model by developing such theoretical elaborations as time lags and non-linear functional forms. Several investigators have attempted to derive improved empirical performance from the model through application of an increasingly sophisticated array of statistical estimation techniques. These efforts are directed toward wringing a more satisfying application of the Basic Model from the data of arms accumulation, the one by tinkering with the theoretical specification of the model and the other by a search for improved parameter estimation techniques. In contrast, others have responded to the unsatisfying performance of the Basic Model by taking greater liberties with the model and the method. One approach has been to test competing theoretical formulations such as organizational, budgetary, and other models focusing on the role of internal or autonomous factors in the accumulation of armaments. A more radical response in the same spirit involves attempts to apply entirely different mathematical structures, generally involving ever-increasing complexity and sophistication, to the problem of arms races. We will consider these various responses in turn.

2.2.1 Beyond Models of Mechanical Arms Accumulation

A frequent criticism of the Basic Model relates to its "mechanical" nature. The model describes a continuous and instantaneous mutual reaction process, but it gives no clues as to how or why this process should occur. Attempts to confront this problem have generally taken one of two lines of approach: Change the specification of the model or implement the ideas of the model using a different mathematical formalism. The former tendency is usually expressed by attempts to incorporate time-lags into the Basic Model and by attempting to specify non-linear formulations of the model, often pushing beyond the bounds of mathematical tractability. The latter approach includes the use of "graphical statistics," geometrical, and various stochastic methods. Both approaches advanced arms accumulation modeling at the margin.

Some attention has been focused on the instantaneous aspects of the Basic Model, that is, on the lack of explicit representation of effects of the past on behavior in arms accumulation. The idea here is that reaction processes are perhaps not instantaneous but rather involve a time delay of one year (or two or three . . .) during which the leaders and/or bureaucracies of a nation develop policy responses to changes in the armaments behavior of others. One of the first attempts to confront this difficulty was Abelson's respecification of the Basic Model to include a "memory" of past provocations.[46] This foreshadowed a larger set of attempts to improve the model by incorporating time-lags.[47]

Another response to the inadequate empirical performance of the Basic Model focuses on its linear functional form. A look at the empirical data of arms accumulation often strongly suggests that the process is far from linear in form. Much of the textual and polemic literature suggesting that arms races lead to an explosive spiral of accumulation patterns would be nonsensical if the process were indeed linear. R.L. Hamblin and others compared the Basic Model to non-linear models in seven arms races, finding the non-linear form generally superior in performance.[48] Irwin Sandberg, in a theoretical piece, considered the properties of non-linear representations of n-nation arms races.[49] Michael Wallace and Judy Wilson tested a form of the Basic Model modified to incorporate discontinuities and non-linearities on pre-World War I and U.S.-Soviet data, finding relatively good fits; they expanded the model to accommodate multiple nations.[50] In general, however, non-linear difference and differential equation models rapidly approach the boundaries of mathematical tractability and hence seem unlikely to receive extensive empirical consideration in the future.

Partially in response to the intractability issue, many investigators have opted to apply mathematical structures involving probability, geometry, and inequality formulations. Such efforts usually abandon the Basic Model. Of the probabilistic, geometric, and inequality models, the latter have received the least attention.[51] More common are applications of geometry and graphical techniques. These fall into the categories of data presentation, manual curve fitting, analytical geometry, and miscellaneous or unusual applications of graphing techniques. Presentation of data using graphs is a device common to many quantitative efforts, though some employ graphs exclusively for this purpose. Examples include work by George Rathjens, Edgar Bottome, Raymond Hutchings, Klaus Gantzel and Albert Wohlstetter.[52] Prior to the appropriation of more sophisticated statistical curve fitting techniques by the discipline, some authors sought to test agreement between propositions and data by crude manual curve fitting. This approach is seen in work by Smoker, Norman Alcock and Keith Lowe, and Alcock.[53]

The most common use of geometry in the Richardson Tradition involves various applications of analytical geometry to explore weapons state spaces. Among these are some of the earliest attempts to investigate properties of the Richardson models.[54] A general trend toward more sophisticated uses of geometric devices is seen in the work of Mats Friberg and Dan Jonsson, Peter Busch, Knut Midgaard, H. Afheldt and P. Sonntag, Earl Brubaker, Jean-Christian Lambelet, and Stephen Brams and others, culminating in 1982 with Robert Muncaster and Zinnes' vortex theory of internation hostility and war.[55] Bringing up the rear are some unusual applications of geometry, such as the work by J.G. Dash on analogies between magnetism and war.[56]

Among all of the innovative applications of mathematical structures to international conflict, perhaps stochastic approaches have enjoyed the greatest vogue. The stochastic models have explored the applicability of a variety of probability distributions (binomial, poisson, yule, weibull, markov, gauss, diffusion, and contagion) to conflict analysis. Models based on expected utility, expected value, and preference curve formulations have also been posited. Several authors have followed Richardson's interest[57] in probability models of the distribution of wars in time.[58] Others have gone along a similar avenue to explore the spread of conflict in space.[59] Another "natural" application of probability theory to conflict analysis, though derivative of the Richardson Tradition rather than a mainline concern, has to do with the formation of international alliances.[60]

The stochastic approaches outlined above are the main area of interest, but there have been several additional application areas. Models treating the stability of nuclear deterrence constitute another focus for stochastic models of international conflict.[61] In the rival field concerned with Lanchester models of combat, stochastic models are also frequently employed; good surveys of this literature are to be found in texts by L. Dolanksy and Martin McGuire.[62] Other various uses of probability approaches are exemplified in work by Brams, Luc Anselin, Cioffi-Revilla, Piet Rietveld, Mike Horn, and Majeski.[63]

The probabilistic, geometric, and inequality models mentioned here represented a major set of efforts to come to terms with the unsatisfactory performance of the Basic Model of armaments accumulation. Nonetheless, they seem to share some of the shortcomings of the Basic Model. Most particularly, they do not manage to overcome the principal criticism of difference and differential equation forms of the Basic Model. These alternative mathematical structures exhibit the mechanistic character of the original formulations. They give no indication of what motivates the participants in an arms race.[64]

2.2.2 Models of Autonomous Armaments Accumulation

The drive to falsify theories--a signal characteristic of the intellectual activity called science, according to many philosophers of science--has been expressed in the Richardson Tradition in several phases. In the first phase, explorations of the Basic Model often appeared more as attempts to confirm the theory than to falsify it, though measures of correlation and statistical significance provided criteria for rejection of the model. A close consideration of this first phase in the falsification process shows that the results left workers in the Richardson Tradition unsatisfied. The first phase of falsification then dissolved into a variety of competing theoretical formulations focusing on notions of "autonomous" or self-stimulating arms accumulation.

Owing to the generally poor record of the Basic Model when subjected to statistical testing, the Richardson Tradition entered a second phase of falsification efforts. In this second falsification, competing theoretical formulations have been posited and tested against the Basic Model. This was the beginning of a direct challenge to the central touchstone of the Richardson Tradition, the Doctrine of Mutuality. The organizational process models, budgetary incrementalism models, domestic political economy models, and other models focus on ideas about internal reasons for arms accumulation. These autonomy models have received considerable attention.

The typical work in this second phase focused on bureaucracy and economy. Charles Lucier tested an organizational process model on British-Japanese-American behavior in the interwar period.[65] Miraslav Nincic and Cusack explored a model based on the political-economic business cycle.[66] W. Ladd Hollist and Thomas Johnson developed a domestic political economy model including a GNP index and a military research and development component, applying it to U.S.-Soviet competition.[67] These studies are typical of the perspective in the second phase of the Richardson Tradition. A recent critique of myths of military industrial complexes reflects the underlying ideas informing this perspective:

> It may be that the military on either side is engaged not so much in an arms race as in simply doing what it wants to do for its own institutional reasons. The other side is relevant only in that it serves as a convenient excuse for these unilateral activities. Traditional rationales for the arms race present a more straightforward view of military build-ups on either side. We are accustomed to the idea that one side develops a weapons system which is then countered by the other side, leading to a response and so on in an upward spiral of sophistication and cost. . . . Almost invariably, the development of new a weapon or an expansion in production is justified as being simply a response to some initiative on the enemy's part. But the record indicates that the desire for the new weapon or a longer production line comes first; only afterward is the threat discovered that the new weapon is supposed to meet.[68]

Beyond the unsatisfactory empirical performance exhibited by the Basic Model, a significant motivation underlying the drive to examine models based on autonomous factors had to do with the mechanical character of the Basic Model. If the formally content-free concept of interaction embodied in Richardson's warpeace particle left something to be desired when it comes to the interpretation of defense coefficients, it was thought, then perhaps an autonomous model might yield not only improved empirical performance but also a more convincing substantive interpretation for the parameters of the model. As budgetary incrementalism, as policy momentum or as macro-economic influences upon governmental expenditures, interpretation of the parameters in autonomous models offered firmer footing on a field rich in explanatory possibilities.

Hollist, Willem Saris and Cees Middendorp, and Wallace all compared internal and external models. Hollist's effort is the most comprehensive.[69] He tested eight model variations on data from U.S.-Soviet, Israel-Egypt, Iraq-Iran, and India-Pakistan conflicts, finding that explanations focusing on internal factors are superior. Working with data from U.S.-Soviet competition in missile procurement, Saris and Middendorp's model of production capacity, technological innovation, and force level goals performed better than the Basic Model.[70] Only Wallace found greater support for reactive-type models over internal models.[71]

Head-to-head tests of autonomy and mutuality models are perhaps best exemplified in the work of Charles Ostrom.[72] In his first excursion in this direction, Ostrom tested a Basic Model against an organizational process model using U.S. arms expenditure data. His conclusion that neither the autonomous nor the mutual formulations "can be identified as relatively accurate" was later refined to a "reactive-linkage" model combining elements of both the reaction process and the organizational process explanations. Again using U.S. data (1954-1973), the synthesized reactive linkage model provided a superior fit compared to models relying solely on either the autonomy or mutuality formulations. Following Ostrom and J.L. Stromberg,[73] Majeski's thesis extended the idea of reactive-linkage in the form of a multi-level disaggregated decision-making model.[74] Similarly, Gernot Kohler developed a synthesis of Richardson's Basic Model and the *autisme* model offered by Dieter Senghaas,[75] referring to the result as a "structuro-dynamique" model.[76] Zinnes also compared interactive and autoregressive models, settling on a combined model.[77] These early combined models foreshadowed the next phase of investigation in the Richardson Tradition.

2.2.3 Mixed Models of Arms Accumulation

Given the general thrust of the findings in the second phase of falsification, numerous workers moved into a third phase of falsification, signifying a major turning point in the epistemological history of the Richardson Tradition. In this third phase, models based on the notion of autonomy have been integrated with models of mutuality, echoing Richardson's movement from the elementary to the basic model. Among the first to consider explicitly integrating the internal and external models was Gantzel, who suggested that East-West conflict has been characterized by a progression through "inner-directed" and "other-directed" periods.[78] Raymond Tanter and William Potter also considered East-West conflict, devising a model that combines interactive and organizational process inputs.[79] This approach predates several later efforts relying on the organizational theories of Graham Allison,[80] who himself explicitly challenged the notion of formally content-free action-reaction processes in armaments competition.[81] A follow-on study by Allison and F.A. Morris strengthened the belief of these authors that internal factors predominate over external ones in weapons procurement decisions;[82] they directly challenged ideas on arms race action-reaction put forward in the work of Rathjens.[83] Also taking cues from Allison was Michael Dennis, who examined the naval procurement policies of France, the United Kingdom, the United States, and the Soviet Union between 1950 and 1970, finding that intra-service perception is an important factor in arms accumulation processes.[84]

Thus, the early challenges to the Doctrine of Mutuality led through a series of phases in which autonomous formulations were tested against the Basic Model of armaments accumulation, eventually leading to a synthesis of mutuality and autonomy models. While the empirical performance of mixed models has not been so impressive as to set the matter to rest, the emergence of a set of explanations moving away from the abstract and content-free mechanistic interaction models advanced the state of the art, presaging the next major phase in the epistemological history of the Richardson Tradition.

2.2.4 Models of Teleological Accumulation

Anatol Rapoport played a central role in the dissemination and popularization of Richardson's work. Rapoport, founder of the *Journal of Conflict Resolution*, devoted an entire issue of the first volume of that journal to a critical review of Richardson and his theory. The emphasis should perhaps be placed on the adjective, "critical," for while Rapoport lauded Richardson's work as the "embodiment of the best in scientific ethics," he expressed strong reservations concerning the applicability of Richardson's methodology to the study of human behavior. "Classical physics postulates the operation of blind forces on the 'here and now' and denies to matter any imagination, any scanning of alternatives, any choice, or any foresight. Richardson's method is essentially physical."[85] Contrasting the determinism of Richardson's "social physics" with the indeterminate, choice-laden finite mathematics of game theory, Rapoport presciently wondered whether physics failed social science by giving too little of a role to the rational elements of human social behavior, and game theory, too great a role: "Possibly the true nature of large-scale human events is intermediate between physical determinism and cognizant choice based on evaluation of potentialities. If so, then the findings of game theory and systems of differential equations are the two extremes bracketing the as yet unknown theoretical method suitable for the study of human behavior."[86] Though Rapoport's review of Richardson was the first to appear in a political science journal, the themes he sounded and the avenues of exploration he proposed still reverberate in the tradition today.

Ten years after Rapoport introduced the discipline to Richardson, critical students of the Richardson Tradition were able to note little in the way of substantive theoretical progress. Hayward Alker, Jr., seemed to be particularly galled. "Is it not in some way ironic that after Richardson and Rapoport developed and expounded models of international behaviors 'where the participants do not stop to think' with the intent of changing these patterns," Alker inquired in his inimitable Socratic fashion, "that so many mechanistic, nonteleological, nonadaptive, non self-organizing [*sic*] models of arms races are continually brought to our attention by Richardson and others?"[87] Laboring under the influence of Talcott Parsons, Alker reminds us that central to the business of social action and social science is the concept of values and, more to the point, the consequences of the values to others. There is no concept of values attached to the Basic Model, no notion of purposive action or adaptive behavior, none of goal-directed cognition nor of the dynamic behavior exchange driving the adaptation of the structures underlying social reality.

These concerns motivated a fourth phase of inquiry in the Richardson Tradition, the search for teleological models of armaments accumulation. At the crest of this fourth wave were three teams of researchers: Brito and Intriligator, Simaan and Cruz, and Gillespie and Zinnes. Gillespie and Zinnes expressed the dominant thrust of this phase most directly: ". . . as Rapoport, Alker and others have pointed out, the [Richardson Basic] model is a-rational in the sense that it does not explain how nations come to adopt their particular reaction equations. Thus to answer critics like Rapoport and Alker it is necessary to extend the Richardson model in such a way that a decision calculus involving national goals becomes a part of the theoretical structure."[88] Rather than appropriating an "as yet unknown" method lying between the extremes of game theory and differential equations as Rapoport speculated, however, the authors of the fourth phase instead synthesized these extremes with theories of optimal control and differential games.

Let us suppose for a moment that there exists an armaments accumulation situation involving two nations, one that is described by Richardson's Basic Model. Further, suppose that there exist several different "equilibrium levels" of armaments expenditures, i.e., levels of arms spending at which the system exhibits stability such that following perturbations the system returns to a normal or "steady" state. How can we identify the minimum amount of arms expenditures required to achieve equilibrium? This is precisely the type of problem that optimal control theory was designed to deal with.

"Optimal control theory assumes we know the basic cause-and-effect relations in the empirical world. These empirical generalizations are called the 'plant' or the 'system.'" The Basic Model (or some variation) represents the plant and, clearly, that the plant is known is a strong assumption in the Richardson Tradition. In any case, as Gillespie suggests, "The goal of optimal control thinking is the answer to the question 'Given the plant, how can the goal or objective be satisfied?' With appropriate mathematical representations, this question can be answered; it will yield the best strategy or action for a decision maker to follow in order to arrive at his objective."[89]

Perhaps the first application of mathematical optimization to problems of armaments accumulation modeling was Alain Enthoven's early work at Rand, as reported by C.J. Hitch and R.N. McKean.[90] Another important early effort is found in Intriligator's work on nuclear targeting and rates of fire.[91] P.E. Chase's contributions on feedback control theory were also a precursor to this line of inquiry.[92] The principal technically competent applications of optimal control theory in the Richardson Tradition have been made by Simaan and Cruz, Brito and Intriligator, Gillespie, Zinnes, G.S. Tahim, and Philip Schrodt.[93] Other applications of mathematical optimization methods to arms accumulation and related problems include Robert Kupperman and Harvey Smith on theorems about deterrence,[94] Martin Sampson's concept of the "policy zone,"[95] William Baugh's transient response analysis of Richardson equations,[96] Gilmer Blankenship on population dynamics,[97] and Warren Phillips on C^3I (command, control, communications, and intelligence).[98]

Beyond the questionable assumptions upon which the optimal control approach depends, there is an additional difficulty with this approach. The mathematical underpinnings of the optimal control approach to arms race modeling are the most sophisticated mathematical structures that have yet been applied to the problem. This relatively high degree of mathematical complexity carries with it a serious cost. While the technical sophistication of the social science community has grown impressively in the last quarter century, it remains true that only a very few social scientists possess the skills required to evaluate and refine the optimal control approach. This carries the danger that the few committed to this approach have or will become a micro-tradition, quietly practicing their art for an audience of half a dozen like-minded scholars, while leaving their thousands of less mathematically inclined colleagues to wonder who was Pontriagin and what Hamiltonian functions have to do with war and politics. "If they want to maximize their contribution to understanding," as James Rosenau argues, those who use such mathematical methods "will have to maintain and extend their contacts with the much larger community of non-mathematical students of world politics, a task that requires them to recall constantly that their mode of inquiry is a form of language that needs to be translated if their ideas and findings are to circulate widely."[99] Compounding the challenge is the fact that for many of the more advanced mathematical techniques, there simply exists no natural language analogues with which to express and translate the mathematical concepts. Yet despite these difficulties, as Gillespie and Zinnes point out, applications of optimal control theory to the study of arms races did represent a substantial advance in the art because it allowed the introduction of elements of choice to the specification of models in the Richardson Tradition.[100] In order to push the introduction of elements of choice farther yet, many of the optimal control analyses mentioned above also integrate ideas from game theory.

Game theory provides a systematic method for the analysis of choice in situations with conflict of interest. Distinguishing games of chance, games of skill, and games of strategy, game theory concerns itself only with the latter.[101] The general objective in game theoretical analysis is to identify a choice or series of choices (i.e., a strategy) that yields the highest payoff in a given game and to determine if that game has a "stable" solution (i.e., one that will always be chosen by rational players because it contains their respective optimal strategy).

Of particular interest to the study of international relations is a class of games of strategy known as two-person non-zero sum games. In this kind of game, players have both conflicting and common interests, as is typical of many situations found in international relations. In fact, a two-nation arms race is often used to illustrate of one such game, the "Prisoner's Dilemma."[102] The problem in non-zero sum games like the Prisoner's Dilemma is that solutions are always problematic. Unlike zero-sum games, for which the Minimax Theorem assures a solution,[103] non-zero sum games do not necessarily have a stable equilibrium solution, and, even if they do, we have no completely convincing approach for finding the solution(s). In other words, these are games where it is likely that everyone will lose in the end.

Several approaches to finding an equilibrium solution in non-zero sum games have been put forward, though, and one that has found application to arms accumulation was suggested by Nash.[104] We need not be concerned here with the technical details of the so-called Nash Equilibrium, but the general idea Nash demonstrated is that there may exist a solution such that no choice available to the opponent can unilaterally increase his payoff or decrease yours, though if he deviates from the solution he may decrease his payoff. Such a situation is obviously stable when both players are fully informed and completely rational. Simaan and Cruz were the first to note the applicability of the Nash Equilibrium to problems of the Richardson Tradition:

> . . . if it is possible for the [closed-loop Nash] optimized system to take the same structure as Richardson's model, then this approach would give the decision-makers a procedure for computing the unknown coefficients . . . in order for them to compute their armaments budgets for any given period of time. These budgets so computed will ensure a state of equilibrium between the two nations in the sense that if each nation holds fast to its budget, the other nation will only hurt itself if it tries to deviate from its Nash budget.[105]

The three teams of Simaan and Cruz, Brito and Intriligator, and Gillespie and Zinnes have done the principal work in this area.[106]

There are two principal criticisms of the efforts employing optimizing game models with Nash Equilibrium solutions. In the first place, as Simaan and Cruz point out, the existence of a Nash solution is not guaranteed.[107] A second and more crippling problem with this work flows from the assumptions Nash had to make in order to achieve his solution. He assumes (1) both players have full knowledge; (2) full estimation of cardinalized utility functions; and (3) optimizing choice.[108] It has been argued that these assumptions constitute a highly unrealistic representation of complex decision processes.[109] In view of the evolving literature on decision making and organizational behavior from James March and Herbert Simon and R. Cyert and March, to David Braybrooke and Charles Lindbloom, and Simon, through John Steinbrunner, Anselm Strauss, and Simon, and finally David Maines, Richard O'Toole and Anita O'Toole, and Simon, it is difficult to disagree with this criticism.[110] Teleological models of armaments accumulation, the culmination of the fourth phase of falsification in the Richardson Tradition, remain unequal to the explanatory performance scholars have demanded.

2.3 The Puzzle Remains

As evidenced in the foregoing literature review, we have seen that the history of theorizing in the Richardson Tradition of arms race analysis has been a series of applications of various quantitative mathematical structures to the subject matter of armaments accumulation, both theoretical and empirical. In Thomas Kuhn's terms this can be viewed as the evolution of a scientific paradigm, appearing as a series of "Is It Like . . . ?" inquiries.[111] Is it like a coupled system of positive and negative feedback inputs? Is it like a system in exponential decay? Is it like a wave? Is it like a lag structure? Is it like a differential game? Is it like a Poisson stochastic distribution? This is a classic strategy from the physical sciences.

Richardson's attempt to deploy the methodology of the physical sciences directly to the subject matter of the social sciences can rightly be referred to as a "social physics." As Rapoport notes, "the name is more than a metaphor. The method is actually the method of theoretical physics."[112] My account of Richardson's training and the parallels between his contributions to the physical and social sciences confirms this. Despite, or perhaps because of the elusiveness of satisfying findings within the Richardson Tradition, there is a continuing impulse to seek new mathematical structures in a search for the answer. Yet some have urged the tradition to adopt a more cautious approach. They argue that "much of our surrogate modeling with physical science techniques would benefit by grinding to a halt as in itself, it has not magic . . . understanding without foundation is worse than ignorance."[113]

2.3.1 Equivalence of the Mutual and Autonomous Models

Two of the principal axes of theoretical investigations into armaments accumulation processes have dealt with "mutual" or interactively driven processes and "autonomous" or internally driven processes. Attempts to disentangle the possible combinations of these effects in empirical investigations of armaments accumulation processes, as the foregoing literature review suggested, have mostly come to grief. "The common failure of statistical tests to discriminate among various models indicates that there are fundamental problems with either the explanatory power of the models or the discriminating ability of the tests. Which is the real difficulty remains a subject for future research."[114] The present research can suggest some possible considerations toward a determination on this question.

It can be shown that the common algebraic representation of the general form of the mutual arms process model,

$$X(t) = a_1Y(t-1) + a_2X(t-1) \qquad [2.12]$$

$$Y(t) = B_1X(t-1) + B_2Y(t-1) \qquad [2.13]$$

is formally equivalent to the common algebraic representation of the general form of the autonomous process model,

$$X(t) = X(t-1) + a_3X(t-2) \qquad [2.14]$$

$$Y(t) = Y(t-1) + B_3Y(t-2) \qquad [2.15]$$

This result is well known for the particular case, following work by Schrodt, Zinnes, Gillespie and Tahim, and Squires.[115] For the general case, if one transforms the mutual process model to the second order form (i.e., lag Equations 2.12 and 2.13 by one period) so that t --> t-1, then

$$X(t\text{-}1) = a_1Y(t\text{-}2) + a_2X(t\text{-}2) \qquad [2.16]$$

$$Y(t\text{-}1) = B_1X(t\text{-}2) + B_2Y(t\text{-}2) \qquad [2.17]$$

Substituting 2.16 into 2.12 and 2.17 into 2.13 gives

$$X(t) = a_1(B_1X(t\text{-}2) + B_2Y(t\text{-}2)) + a_2X(t\text{-}2) \qquad [2.18]$$

$$Y(t) = B_1(a_1Y(t\text{-}2) + a_2X(t\text{-}2)) + B_2Y(t\text{-}2) \qquad [2.19]$$

Then rearranging the terms of 2.12 and 2.13, one notices that

$$Y(t\text{-}1) = (X(t) - a_2X(t\text{-}1))(1/a_1) \qquad [2.20]$$

$$X(t\text{-}1) = (Y(t) - B_2Y(t\text{-}1))(1/B_1) \qquad [2.21]$$

Lagging the terms of 2.20 and 2.21 by one period to obtain

$$Y(t\text{-}2) = (X(t\text{-}1) - a_2X(t\text{-}2))(1/a_1) \qquad [2.22]$$

$$X(t\text{-}2) = (Y(t\text{-}1) - B_2Y(t\text{-}2))(1/B_1) \qquad [2.23]$$

now 2.22 and 2.23 can be substituted into 2.18 and 2.19 to yield

$$X(t) = a_1(B_1X(t\text{-}2) + (1/a_1)(X(t\text{-}1) - a_2X(t\text{-}2))(B_2)) + a_2X(t\text{-}2) \qquad [2.24]$$

$$Y(t) = B_1(a_1Y(t\text{-}2) + (1/B_1)(Y(t\text{-}2) - B_2Y(t\text{-}2))(a_2)) + B_2Y(t\text{-}2) \qquad [2.25]$$

Simplifying 2.24 and 2.25 we finally obtain

$$X(t) = X(t\text{-}1) + a_3X(t\text{-}2) \qquad [2.26]$$

$$Y(t) = Y(t\text{-}1) + B_3Y(t\text{-}2) \qquad [2.27]$$

or in other words exactly the general form of the autonomous model in Equations 2.14 and 2.15. Thus, it is proven that the common algebraic representations for the external reaction process or mutuality model and the internal organizational process or autonomy model of armaments accumulation are exactly the same thing.

While this result is well known for the particular case, its implications are not much discussed in the literature. One might ask what the implications of this equivalence are in the general case. Robert Huckfeldt, C.W. Kohfeld, and Thomas Likens take a similar approach to derive a similar result.[116] They suggest only that the explicit theoretical interaction expressed in the mutuality model is transformed so that the interaction is implicit in the increased order of the lag structure in the recursive form of the model. This seems to stretch the bounds of interpretation past the point where meaningful interpretation is possible. How are we to know which model represents which theoretical idea of the underlying process? Just because the investigator says that this parameter represents country "A" (for the internal case) or country "B" for the external case?

Let us invent an imaginary mathematician for a professional opinion. The mathematician may object to this entire discussion. What does it matter that the class of equations under consideration contains several ways to express a given data vector? This is precisely the power of the mathematics! Both forms have exactly equivalent mathematical meanings, as two different ways to say the same thing. It's purely a question of interpretation. In the same way, equations of the same form can represent a spring or a pendulum in mechanics or a negative feedback system in electronic circuits. This is so because the equations themselves are not "about" any of these particular things. The equations are formally content-free abstractions and refer to nothing outside of themselves. If it is disturbing that internal conceptions of armaments accumulation processes appear to have precisely the same mathematical form as conceptions of a supposedly different process, an external armaments accumulation process, then the problem is not with the mathematics, it is with the level of abstraction of the conceptualization and the "granularity" of the syntax chosen to represent the descriptions of the two processes. A data series that is increasing in a smoothed exponential process can be represented by an infinite number of different equations.

This of course points right to one of the main difficulties with the differential equation approach. Such equations are general descriptions of general processes and as such are very generalized tools, more like a shovel than a scalpel. Interpretation of the parameters, though very much an "art," is critical in an empirical science where we need to know which interpretation on the meanings of the parameters is correct.

What this equivalence means for theory is that we have not been able to disentangle the effects of mutuality and autonomy in arms accumulation. The syntax and granularity of the formalisms used to model mutuality and autonomy are such that the two different ideas are represented by the same mathematical forms. This suggests that the explanatory power of the models in the Richardson Tradition is deficient, unable to discriminate between the various "internal" and "external" effects in an armaments race. It may be necessary, ultimately, to find additional tools in order to solve this particular challenge. But the chief difficulty is not a technical one. As Stoll argues, "most previous work fails to deal directly with the question addressed here: does the fault lie with the conceptualization of the arms acquisition process, rather than with any particular data set or estimation technique?"[117] Stoll means to imply--and one cannot but agree--that the original conceptualization of armaments accumulation processes as a product moment in a formally content-free state space leaves aside some of the elements crucial to the explanation and understanding of arms accumulation. The content-free abstractions borrowed from the physical sciences have not been entirely helpful for systematically examining the problem of intentionality.

2.4 They Did Stop to Think--About Strategy

It has been part of the American character to be suspicious of "strategy" and "strategic thinkers" at least since President Washington warned against entanglements with the politics of the old world. Treated since antiquity, the concept of strategy evokes the words and deeds of Alexander the Great, Thucydides, Sun Tzu, Tacitus, Machiavelli, Napoleon, Hitler, Ghandi, and other evil geniuses and saints. Scientific treatments of the concept, in bald contrast to the practical kind, are usually dated no earlier than Antoine Augustin Cournot and Blaise Pascal, and sometimes as late as the early work of Emile Borel and John Von Neumann.[118] There is a distinct tension between the theory and the practice of strategy. Though the concept of strategy is central to the business of armaments accumulation, rigorous examination of the concept is inhibited by this tension between theory and practice, no less than by the prejudices of the American character.

In this part, I consider the concept of strategy. A strategy will be defined here as a political action plan. Several alternative denotations of the term help to inform a conceptual framework for thinking about strategy. In the real world, many view the role of strategy in armaments accumulation simply as war by other means.

2.4.1 Arms Accumulation as War by Other Means

When former President Richard Nixon published his book, *The Real War*, it was not surprising to find that he thinks we are at war with the Soviet Union.[119] Indeed, Nixon's thesis was hardly revolutionary. His point is one that political philosophers have repeated since the dawn of history. As Thomas Hobbes reminds us in Chapter Thirteen of *The Leviathan*, "war consisteth not in battle only, or the act of fighting": "For as the nature of foul weather lieth not in a shower or two of rain, but in an inclination thereto of many days together: so the nature of war consisteth not in actual fighting, but in the known disposition thereto during all the time there is no assurance to the contrary." There is "no assurance to the contrary" in a system of states. Treaties and armistices notwithstanding, the Sovereign is in a perpetual State of Nature, which is to say, War.

Although one can never prove a case by any number of examples, an illustration is instructive. Americans are socialized to believe that the United States is a peaceful and non-aggressive nation. Prudence dictates, as Plato recommended, that some aspects of politics and history be withheld from young ears. The recent history of war planning in the U.S. government is an example of this, suggesting, as it does, that there is something more to "peace" than simply goodwill.

In 1941, even as the darkest days of World War II lie ahead--indeed, even before Pearl Harbor--General Marshall ordered U.S. war planners to look past the end of the war, to envision the next war, and to prepare plans for it.[120] At first, planners in the Army Air Force (AAF) Post War Division thought only of a resurgent Japan or Germany, but very soon, prior to the fall of Germany, plans were afoot to attack the U.S.S.R.[121] AAF planners by early 1945 had developed a strategy that entailed air attacks on twenty key Soviet cities, modeled after the campaign against the German petroleum industry.[122] The events of August 1945 suggested a significant new twist, and within a matter of months--by June 1946 --planners of the Joint Chiefs of Staff had developed the war plan code-named PINCHER, calling for a strategic air offensive against twenty-four Soviet cities using nuclear weapons.[123]

The target list has grown inexorably since that time, into today's "SIOP," which defines untold thousands of people, places, and things thought to be in need of atomic bombardment in the event of overt conflict. The huge list of items the United States is prepared to vaporize--it is among the most sensitive of defense secrets, labeled with innocent acronyms like JSTL (Joint Strategic Target List)--is constantly updated to reflect the latest in doctrinal nuance and technical achievement. One cannot overlook that fact that ideas like the JSTL are among the very nastiest business human beings have ever conceived. Though details are lacking, it is a reasonable certainty that the Soviet Union holds a strikingly similar plan, differing principally in the target coordinates.

There can be no doubt that this conforms to Hobbes' "known disposition to fighting." We are at war. We always have been. The difficult part to come to terms with is that, according to philosophers, this is a natural state of affairs. Planning wars is the normal business of military bureaucracies; this is what they do, their purpose. The rationalizations are all complete; the weapons aimed and cocked; the generals on station around the clock; the "Football" always by the leader's side. U.S. and Soviet air defense forces intercept each other's training and reconnaissance sorties, U.S. and Soviet naval vessels play tag across and under the seven seas, and U.S. and Soviet soldiers stare at each other through gun sights across frontiers from Berlin to Panmounjon. While they cannot see into the enemy's mind, to divine his plans and intentions, they can see down the barrel of his gun. All that is lacking is an occasion.

This discussion highlights one contention about arms races. Much debate in the literature over whether or not arms races lead to war is, simply, conceptually unclear. Arms races are a form of war. Such questions could be better posed as, say, What are the conditions under which such plans progress to fighting, when the ideas give way to actions and the destructive potential becomes actual?

2.4.2 Conceptions of Strategy

Up to this point in this work--as in much of the literature on arms races--the term *strategy* has been used rather loosely, without making much of an effort to be clear about the meaning and scope of the term. It has been used as as a synonym for the more general idea of a plan, with obvious political and military connotations. Yet strategy is a concept over which much ink has been spilled and to which many meanings and implications have become stuck fast. With regard to the definition of strategy, it is perhaps chastening to observe that even elite military leaders are sometimes hard pressed to attain conceptual clarity on this seemingly central idea. For example, after a recent conference on the concept of strategy, cosponsored by the National Defense University and the Office of the Assistant Secretary of Defense for International Security Affairs, an organizer confessed that "it is clear that no simple definition adequately addresses the complexity" involved in considerations of national strategies.[124] To be clear, nonetheless, it is necessary to venture some distinctions.

Despite the complexities, a working definition of strategy should be offered. A strategy will be defined here as a *political action plan.* A plan is "any hierarchical process in the organism that can control the order in which a sequence of operations is to be performed."[125] The "political" will be defined following Harold Lasswell and David Easton.[126] A strategy is therefore a hierarchical process that aims to control the authoritative allocation of values, or who gets what when and how. Strategies, it will be noted, evolve and unfold in multiple overlapping contexts. Simply stated, then, a strategy is a plan to get what you want. Simplicity tends to reveal fundamentals: What you want is for your values and intentions to prevail.

This is consistent with the idea of strategy as conceived in game theory models of armaments accumulation processes, as we have seen. Briefly, this formal notion of games of strategy from game theory is concerned with finding stable equilibria in systems involving competing values. A serious problem with the game theoretical notions of strategy was mentioned, *viz*, the indeterminancy of equilibria in non-zero sum mixed games. This problem stems from the fact that in the formal theory of games, one area of insufficient theoretical development is that of cooperative and mixed cooperative-conflict games. We seem to know more about conflict than about cooperation, even though cooperation is manifestly as real and important a phenomenon of social behavior as is conflict. This is but one illustration of the gulf between the theory and the practice of strategy. Another is the postulate of rationality. So different are the scientific and the practical treatments of strategy that the layman might be forgiven should he believe that the two ideas of strategy reference fundamentally different matters.

The meaning of strategy when used by soldiers is also suffused with the notion that one's intentions should prevail, yet the soldiers' use of strategy seems quite different than the scientific sense of the term. Political theories of military strategy, i.e., strategies of conflict, have received extensive treatment at the theoretical level by some of the finest soldiers in each of the great civilizations of history. Comparatively little scientific work, however, has ever been undertaken on these political-military strategies of conflict. This has contributed, perhaps, to the lack of conceptual and theoretical work on how strategy changes. Rosenau's studies of political adaptation are as close to this type of research as anything that comes to mind.[127]

Little or no research has been done on the relationships among grand strategy, military strategy, and operational strategy. The works of two soldiers, Karl Von Clausewitz and B.H. Liddell-Hart, appear to be the most systematic and scientific treatments of these problems in the literature. Virtually no work has been done on how strategic planning in a nation-state changes over time, though it seems to me that this is something fundamental to arms race parameters. One significant recent exception is the innovative and important, but incomplete work by Noel Kaplowitz.[128] He examines the relationships among sets of strategies, considering their interactions and their transformations. This is precisely the sort of inquiry required to elaborate an empirical science of strategic behavior. A premier aim of modeling in arms accumulation research should be the representation and testing of Kaplowitz' kind of strategic transformation structures, as revealed through empirical inquiry.

Table 2.2 is tentatively advanced as a formulation of a strategy state space, useful only in illustrating how difficult are the challenges of organizing this problem in a tractable fashion. The traditional lexicon of strategy in the politico-military context will be discussed in some detail in Chapters Four and Five, but the extent to which these historical, technical concepts mirror the underlying dynamics of the phenomena remains an open and puzzling question. As will become apparent, the problem of intentionality operative here is neither simple nor singular.

What is missing is a theory that describes a state space of strategies. The appropriate metric for such an effort remains elusive, but it will be necessary to find some way for representing the transformations among different strategies if the Richardson Tradition is to make headway with the problem of interaction in armaments accumulation processes. Several formalisms suggest themselves as possible structures to begin representing such concepts. One idea is the augmented transition net[129] as used in the network-based planning system, NOAH;[130] another is the "frame" or "script" representation[131] as used in the POLITICS inferencing system.[132] A recent application of the script concept can be found in Richard Herrmann's interesting proposal for making inferences about the perceptions of strategic actors.[133]

To summarize, conceptions of strategy vary greatly among different authors coming from both scientific and military perspectives. All in all, strategy remains enigmatic and unclear, yet seemingly a central concept. Much basic work remains to be done. About all that seems consistent between scientific and traditional conceptions of strategy is that both seem to boil down to the competition of values and applications of will directed to preferred outcomes. This returns us to our central theme: Intentionality is central to the process of arms accumulation. And a better understanding of strategy and how it evolves lies on the path to better models of arms races.

2.5 Bureaucracy and Arms Accumulation Theory

From our review of the Richardson literature, it is apparent that one consistent theme emerging from the work of the last quarter century is a trend toward incorporating aspects of bureaucratic behavior into theories of armaments accumulation behavior. In fact, this concern is evident in some of the earliest recorded thought on the topic of military interaction. Thomas Schelling, for example, cites Xenophon's discussion from the fourth century B.C. of the effects of the dynamics of bureaucratic interaction between the Greek and Persian military organizations. "I observe that you are watching our moves as though we were enemies, and we, noticing this, are watching yours, too."[134] The problem of organizational interaction between military bureaucracies, then, is an ancient one.

Despite the longevity of this problem and the recent efforts among Richardson's students to approach it more theoretically, the incorporation of ideas about bureaucratic behavior in arms accumulation theory remains underdeveloped. Though much spade work remains to be done in this field, it can be shown that extant advances in the theory of organizational behavior are available for immediate incorporation into our arms accumulation theories. A comparison of the current state of organizational theory and the dominant applications of organizational theory in arms accumulation studies will reveal a disparity that demonstrates this contention.

Table 2.2 A Strategy Space

ALL POSSIBLE STRATEGIES

- STRATEGIES OF CONFLICT
  - TOTALISM
    - Totalist
    - Long-Run
    - Latent
  - MAXIMIZING
    - Competitive
    - Use-of-Force
    - Deterrent-Punitive
      - Military Strategies
        - Air Land Battle
        - Continental Defense
        - Sea Control
        - Invasion Force
        - Power Projection
        - Selective Annihilation
  - SATISFICING
    - Mutual
    - Other
- STRATEGIES OF CONCILIATION
  - terra incognita

Note: Relies heavily on Kaplowitz, 1984

2.5.1 Types of Organizational Decision-Making Theories

Social scientists in this century have produced a welter of theories that attempt to explain the behavior of that ubiquitous modern institution, the large bureaucratic organization. This has helped to make the literature on the topic voluminous and somewhat opaque. Much of this literature has been produced by political scientists in efforts to understand the role played by organizations in political life. In an attempt to render the literature more comprehensible and at the risk of some oversimplification, the literature on organizational decision making can be divided into three broad categories. These are the "Rationalist," "Structuralist," and "Interactionist" approaches to the analysis of organizational behavior. These categories correspond closely but not perfectly to those posited by Graham Allison.[135] These three types of models are differentiated according to the particular emphases embedded within the assumptions constituting the theories.

Among the theorists and theories considered to be of the Rationalist type are those who refer to the "Rational Actor,"[136] the "Analytic Paradigm,"[137] the "Nash Strategy,"[138] "Utility Maximizing,"[139] "Rational Choice,"[140] and "Rational" models of decision making.[141] In the Rationalist decision theory, our focus is quickly drawn to the structure of the decision problem itself. The principal analytical activity here revolves about the relationship between the global goals of the organization and the state of the environment. The obstacles and opportunities present in the environment relative to the goals of the organization constitute the "decision problem," and so once these qualities have been defined, then the remaining task is to weigh the alternative choices so as to maximize utility.

Those who are classified as being of the Structuralist type would tend to refer to "Organizational Processes,"[142] "Organizational Structure,"[143] "Bounded Rationality,"[144] "Disjoint Incrementalism,"[145] "Satisficing,"[146] the "Cybernetic Paradigm,"[147] and the "Decision-Making"[148] approaches to the study of organizational behavior. Structuralist decision theory, in contrast the rationalist approach, focuses on the structure of the decision maker or decision-making apparatus. The Structuralist theorist is not so concerned with the decision problem per se but rather with the mechanics of what is to be done about it. Here we tend to concentrate our analytical attention on the rules and regulations or standard operating procedure (SOP) of the organization, which controls the flows of information and authority through the hierarchical structure of the organization. The decision problem is no longer viewed as a unified whole, which is to be decided instantaneously and acted upon as a unit, as in the Rationalist conception. Now the decision problem is decomposed as a series of sub-problems to be sequentially and incrementally attacked.

Third, and finally, the theories of organizational behavior classified as Interactionist are those that refer to "Bureaucratic Politics,"[149] "Human Relations,"[150] "Muddling Through,"[151] "Governmental Politics,"[152] "Artificial Behavior,"[153] and "Symbolic Interactionism."[154] In the Interactionist varieties of decision theory, the focus is drawn to the process of decision making. The decision problem facing the organization as a whole now fades further into the shadows, and one is confronted with a welter of disparate sub-goals associated with the micro-units of the organization, the people. The Interactionist analyst thinks in terms of roles, social conflicts, legitimacy, motivations, fears, hopes, irrational and non-rational constraints--in short--with the messy business of how people "really" interact in the organizational environment.

The Rationalist, Structuralist, and Interactionist types of decision and organizational behavior theory focus on radically different aspects of the whole animal. The divergent assumptions constituting the three types of theory are displayed in Table 2.3. The different types of decision theory have substantial implications for the conceptual adequacy of models of arms accumulation. It is to a discussion of some of these implications that we will now direct our attention.

2.5.2 Rationalist Conceptions Dominate Arms Theory

As we have seen, one of the principal criticisms of Richardson's Basic Model concerned its lack of an explicit rational component. While less insightful critics viewed the Basic Model as a paradigm of irrational behavior, Rapoport, Alker, Gillespie, Zinnes, and others have correctly argued that the original concept is in fact arational. Efforts to integrate autonomous elements such as the bureaucratic momentum in Ostrom's reactive-linkage model represented some of the first attempts to come to terms with this problem.[155] The integration of optimal control theory and game theory represented a more direct attempt to deal with this criticism.

A variety of approaches to incorporating elements of rational decision behavior are found in the Richardson literature. For example, Ostrom, Schrodt, and Chatterjee all consider models of rational behavior in the context of a Balance of Power regime.[156] Brubaker posits models of rational choice in arms accumulation behavior based on economic factors.[157] Brito and Intriligator attempt several approaches, comparing rational and multi-goal models of SALT behavior,[158] combining bureaucratic and economic dimensions of arms accumulation,[159] and developing "softer" conceptions of heuristic decision rules.[160] John Crecine and Gregory Fischer consider the extent to which organizational processes of resource allocation in the U.S. Department of Defense distort rational decision goals.[161] "Expectation" models with strong rationalist overtones have been developed by several investigators.[162] Lambelet considers the possibility that arms races might play a role in the avoidance of war, also a view containing strong elements of the Rationalist perspective.[163] Roslyn Simowitz offers rational choice models of collective goods in alliance formation.[164] This is a sample of the Rationalist models in the Richardson literature, not exhaustive, but suggestive of the wide variety in applications of this paradigm to the modeling of armaments accumulation behavior.

Considerably less common in the Richardson literature are models employing structuralist conceptions of decision behavior. Among these, several different tests of simple organizational process models against the Basic Model have found the former superior.[165] James McCormick compares a model based on Allison's organizational process paradigm to a WEIS-based events data model,[166] and, as previously noted, Allison also considers organizational process models of arms accumulation.[167] Lucier's study of naval budgets and changes in arms race parameters also employs an organizational process model.[168] One final set of contributions in this area is represented by attempts at disaggregation, both of arms categories[169] and of the decision process itself.[170]

Table 2.3
Typology of Organizational Theories

	RATIONALIST	STRUCTURALIST	INTERACTIONIST
FOCUS OF THE MODEL	Structure of Decision Problem	Structure of Decision Maker	Process of Decision Making
VIEW OF THE DECISION MAKER	Unitary Actor	Hierarchical	Distributed
VIEW OF INFORMATION	Perfect or Full	Focused	Overloaded
LEARNING	Causal	Instrumental	Classical
SEARCH	Comprehensive	Problemistic	Heuristic
MEMORY	Unlimited	Partial	Multiple
PROCESSING CAPACITY	Infinite, Parallel	Finite, Linear	Finite, Parallel
GOAL OR VALUE ATTAINMENT	Maximize Values	Satisfice Values	Preserve Values
GOAL OR VALUE INTEGRATION	Complete	Partial	Multiple
DECISION GOOD	Homogenous Utility	Homogenous Interest	Heterogeneous Interest
CHOICE OF OUTCOMES	Mutually Exclusive & Exhaustive	Non–exclusive, Non–exhaustive	Non–exclusive & Exhaustive
DECISION PROCESSES	Unified, Instantaneous	Convergent, Sequential	Divergent, Episodic
SOCIAL PROCESSES	Irrelevant	Set Parameter	Crucial

It is noteworthy that the literature review uncovered no models of arms accumulation behavior employing Interactionist conceptions of organizational behavior, with the possible and significant exceptions of Majeski's recent work,[171] and some recent efforts by Brito and Intriligator,[172] which begin to incorporate the multiple goal structures and heuristic decision processes characterizing the Interactionist perspective. Moreover, there is little systematic work that attempts to compare the various theories of organizational behavior in order to see which might be a better model of the arms accumulation process. Finally, and of most relevance to the present effort, no efforts are known that attempt to integrate insights from the various strands of organizational theory into a hopefully more useful synthesis.

Recalling Rapoport's critique of game theory models,[173] one is reminded of a serious shortcoming in the Rationalist models, which, as suggested above, predominate in the arms accumulation literature. John Harsanyi reiterates it: "But of course the assumption of perfect rationality customary in game theory again goes too far . . ."[174] These theories of rationality fail to correspond with our empirical observations of the imperfect rationality exhibited by humans in their individual and institutional settings. The excessive attribution of rationality to politico-military leaders and their organizations is seen most clearly in applications of game theory employing the Nash Strategy. In order to envision an equilibrium solution, Nash found it necessary to assume that the players be completely rational. Nash players are assumed to have a clear and consistent goal; to identify all possible choices and to calculate their respective outcomes; to assign a cardinal utility score to each of these outcomes; they are assumed to know not only their own utility functions but also those of their opponents; and they are assumed to always choose their best strategy. In other words, the Nash decision maker is perfectly rational. And yet a full generation has passed since Simon conclusively demonstrated that this is, quite simply, an untenable model of human behavior.[175]

In their review of rational choice models in the arms accumulation literature, Zinnes and Gillespie defend against charges of unrealistic assumptions with the argument "that by using rational choice decision-making the modeler is not claiming that in fact decision makers are rational."[176] Somewhat confusingly, they go on to argue that though the actual causes of decision behavior may well lie elsewhere, "we can understand decisions reached in international affairs following this model." This is a peculiar argument and an important point. If decision makers are not rational, then how can we hope to understand their decisions using models of rational behavior? Schelling, Harsanyi, and others hold that such a procedure may be useful for analyzing the extent to which actual decision behavior deviates from a formal model of strict rationality.[177] We agree that this is a valid aim, given that one's purpose is to compare actual behavior to a theoretical ideal. Our purpose in the present effort, however, is different. We wish to specify a model of actual behavior, or one as close to actual behavior as possible, rather than to compare actual behavior to an ideal model. For our purpose, another approach seems advisable. There exists a variety of options for use of organizational decision-making theories in the study of arms accumulation that do not entail reliance on the formal model of rational choice. We will now turn to a consideration of one such approach involving the synthesis of organizational theories.

2.5.3 The Case for a Synthetic View of Bureaucratic Behavior

Graham Allison's *Essence of Decision* offers an important insight that is often missed by analysts who have drawn on Allison's theory.[178] The literature is replete with attempts to test Allison's three models of organizational behavior, to try to find which of the paradigms is true. This is expressed in the arms accumulation literature, as well as in the many other areas to which Allison's work has been applied. When introduced to this body of theory, students often reflect the same seemingly natural response. The majority view is that they like the Rational models and don't like the Structuralist and Interactionist models on the grounds that the latter are somehow too "messy." Perhaps this reflects a natural human bias to want to see human behavior and to interpret human behavior in strictly rational terms. The reality may be quite different, however.

The point of Allison's work was *not* to find which model is "true." No model is true. Each shows us a different aspect of the whole, and each is more or less useful for a given purpose. It all depends upon what one is trying to find out. Like the three blind men and the elephant, one paradigm looks at the foot of the organizational decision-making process; another looks at the ear; the next at an elbow. All are equally useful if used properly and, likewise, equally misleading when misused. This was precisely the beauty of Allison's contribution and, unfortunately it seems, the most commonly overlooked point. Each model is a different way to look at what is going on. Each focuses on decision making with a different "level of magnification," so to speak a microcosm of Singer's Level of Analysis problem applied to the organization.

Thus, it seems clear that it would be useful to exploit this body of theory more fully, allowing the integration of aspects from all three conceptual levels in models decision-making behavior. Moreover, synthesis of decision-making paradigms in arms models would offer wider options for representing the different levels of strategic planning in military bureaucracies previously discussed. In the final analysis, however, a synthetic view of organizational behavior is worth investigating if for no other reason than that such a perspective may be a more realistic perspective on the decision-making process in organizations. The main obstacle lies in finding a way to represent a model of this process, complex enough to contain all the elements yet still tractable enough for treatment in a systematic fashion. A discussion of some potential solutions to this problem, and some conceptual barriers to those solutions, awaits another chapter.

2.6 Summary and Conclusions on Theories

In this chapter we have reviewed the history of the Richardson Tradition of armaments accumulation studies. Several significant points stand out:

§ *The "Doctrine of Mutuality," which forms the very foundation of ideas about the military relationships between nation-states in the Richardson Tradition, was originally modeled as a formally content-free product moment in an abstract "international space."*

§ *The trend of scientific investigations in the Richardson Tradition for the last twenty-five years has centered on a variety of increasingly sophisticated attempts to move from description of the general case to an empirically based understanding of the particular by specifying various mathematical structures to explain the interaction, or lack thereof, of military bureaucracies.*

§ *The cumulative evidence points to bureaucratic and strategic elements that are very difficult to represent and manipulate in a systematic fashion using the traditional mathematical structures.*

§ *Emancipatory conceptions of strategic interaction remain almost completely undeveloped on the theoretical level, and so it is little surprise to learn that the Richardson Tradition has continued to depend upon static, ahistorical conceptions of strategy.*

§ *Arms accumulation models have yet to incorporate the full range of insights from studies on organizational behavior, and there is a need to develop conceptual perspectives on arms accumulation that incorporate and integrate Rationalist, Structuralist, and Interactionist organization decision-making paradigms.*

Based upon these overall conclusions, it is apparent that much work remains to be done in the Richardson Tradition. These observations suggest that we need particularly to sharpen our theoretical concepts about armaments accumulation patterns and to explore new methodologies for representing the interaction behavior of states in arms races. These are the concerns--the development of adequate conceptual perspectives on and exploration of methodologies for the analysis of arms accumulation behavior--occupying the next two chapters.

Notes

1. Bertrand Russell, War, the Offspring of Fear (London: Union of Democratic Control, 1914).

2. Lewis Fry Richardson, "Threats and Security," in Psychological Factors of Peace and War, ed. T.H. Pear (London: Hutchinson and Co., 1950), p. 225.

3. Gregory Bateson, "Culture Contact and Schismogenisis," Man 35 (1935), pp. 173-183; also Bateson's "The Pattern of an Armaments Race--Part I: An Anthropological Approach," Bulletin of the Atomic Scientists 2 (1946), pp. 10-11; and his "The Pattern of an Armaments Race--Part II: An Analysis of Nationalism," Bulletin of the Atomic Scientists 2 (1946), pp. 26-28; Anatol A. Rapoport, "Lewis Fry Richardson's Mathematical Theory of War," Journal of Conflict Resolution 1 (1957), pp. 249-299; as well as Rapoport's "The Mathematics of Arms Races," in International Politics and Foreign Policy, ed. James N. Rosenau (New York: Free Press, 1961), pp. 492-503; his Fights, Games and Debates (Ann Arbor: University of Michigan Press, 1961); and also his "Mathematical Methods in Theories of International Relations: Expectations, Caveats, and Opportunities," in Mathematical Models in International Relations, ed. Dina A. Zinnes and John V. Gillespie (New York: Praeger, 1976), pp. 10-36; Nicholas Rashevsky, Mathematical Theory and Human Relations: An Approach

to a Mathematical Biology of Social Phenomena (Bloomington, IN: Principa Press, 1947); and also Rashevsky's Mathematical Biology of Social Behavior (Chicago: University of Chicago Press, 1951); Quincy Wright, A Study of War, 2 vols. (Chicago: University of Chicago Press, 1942); Kenneth E. Boulding, Conflict and Defense: A General Theory (New York: Harper and Row, 1962).

4. Richard K. Ashley, "Noticing Pre-paradigmatic Progress," in In Search of Global Patterns, ed. James N. Rosenau (New York: Free Press, 1976), pp. 150-158; Hayward Alker, Jr., "Research Paradigms and Mathematical Politics," revision of a paper presented to the International Political Science Association Roundtable on Quantitative Methods and Political Substance: Toward Better Research Strategies, Mannheim, Germany, 1971.

5. John F. Nash, Jr., "The Bargaining Problem," Econometrica 18 (April 1950), pp. 155-162.

6. E. Gold, "Lewis Fry Richardson," Obituary Notices of Fellows of the Royal Society 9 (1954), p. 218.

7. David Wilkenson, Deadly Quarrels: Lewis F. Richardson and the Statistical Study of War (Berkeley: University of California Press, 1980), p. 3.

8. Stephen A. Richardson, "Lewis Fry Richardson (1881-1953): A Personal Biography," Journal of Conflict Resolution 1 (1957), p. 300.

9. Ibid., p. 301; see also the original in Lewis Fry Richardson, Statistics of Deadly Quarrels (Chicago: Triangle Press, 1960), p. xxviii.

10. Gold, "Lewis Fry Richardson," op. cit.

11. Richardson, Arms and Insecurity, op. cit., p. 18.

12. Georg Henrik Von Wright, Explanation and Understanding (Ithaca, NY: Cornell University Press, 1971).

13. Richardson, Statistics of Deadly Quarrels, op. cit.

14. Lewis Fry Richardson, "Mathematical Psychology of War" (N.p.: British Copyright Libraries, 1919).

15. Lewis Fry Richardson, Generalized Foreign Politics (Cambridge, England: Cambridge University Press, 1939).

16. Richardson, Arms and Insecurity, op. cit.

17. Ibid., p. xix.

18. Richardson, "Mathematical Psychology of War," op. cit.; Equation [2.2] is plotted as Line L on Figure 2.2.

19. Claudio A. Cioffi-Revilla, "Theories of Arms Races and Mathematical Structures in International Relations," paper presented to the joint meeting of the Mid-west Political Science Association and the Mid-west section of the Peace Science Society (International), Chicago, April 19-20, 1979.

20. Charles F. Anderton, "Arms Race Modeling: Categorization and Systematic Analysis," paper presented to the annual meeting of the International Studies Association, Washington, DC, March 5-9, 1985.

21. Richardson, "Generalized Foreign Politics," op. cit., and his Arms and Insecurity, op. cit.; see also Norman Z. Alcock, "Arms Races--A Cause of War?" Peace Research 2 (1970), p. 3; and his "An Empirical Measure of International Threat: Some Preliminary Implications for the Middle East Conflict," Peace Research Society (International) Papers 15 (1970), pp. 51-72; and D.L. Wagner, R.T. Perkins, and Rein Taagepera, "Complete Solution to Richardson's Arms Race Equations," Journal of Peace Science 1 (Spring 1975), pp. 159-172.

22. Richardson, "Generalized Foreign Politics," op. cit., Arms and Insecurity, op. cit., and his "Could an Arms Race End without Fighting?" Nature 167 (September 29, 1951), p. 567; see also Paul L. Smoker, "A Mathematical Study of the Present Arms Race," General Systems 8 (1963), pp. 51-60; as well as Smoker's "Trade, Defense and the Richardson Theory of Arms Races: A Seven Nation Study," Journal of Peace Research 4 (1965), pp. 161-176; and Dina A. Zinnes and John V. Gillespie, "Analysis of Arms Race Models: USA vs. USSR and NATO vs. WTO," in Modeling and Simulation, vol. 4, ed. W.G. Vogt and M.H. Mickle (Pittsburgh: Instrument Society of America, 1973), pp. 145-148.

23. Richardson, Arms and Insecurity, op. cit., p. 35; cf. Richardson, "Mathematical Psychology of War," op. cit.

24. Richardson, "Mathematical Psychology of War, op. cit., "Generalized Foreign Politics," op. cit., and Arms and Insecurity, op. cit.; see also William R. Caspary, "Richardson's Model of Arms Races: Description, Critique and an Alternative Model," International Studies Quarterly 11 (March 1967), pp. 63-88; Dina Zinnes and John Gillespie, "Analysis of Arms Race Models," op. cit.; Partha Chatterjee, "The Equilibrium Theory of Arms Races: Some Extensions," Journal of Peace Research 11 (1974), pp. 203-212; Hans Rattinger, "Armaments, Detente, and Bureaucracy: The Case of the Arms Race in Europe," Journal of Conflict Resolution 19 (December 1975), pp. 571-595; Farid Abolfathi, "Defense Expenditures in the Persian Gulf: Internal, Interstate, and International Factors in the Iraqi-Iranian Arms Race, 1950 to 1969," in Exploring Competitive Armaments Processes, ed. W.L. Hollist (New York: Marcel Dekker, 1978), pp. 99-131.

25. Kenneth E. Boulding, Conflict and Defense: A General Theory, op. cit.; Paul A. Smoker, "On Mathematical Models in Arms Races," Journal of Peace Research 2 (1965), pp. 94-95; Caspary, "Richardson's Model of Arms Races," op. cit.; Mats Friberg and Dan Jonsson, "A Simple War and Armaments Game," Journal of Peace Research 5 (July 1968), pp. 233-247; Murray Wolfson, "A Mathematical Model of the Cold War," Peace Research Society Papers 9 (1968), pp. 107-123; P.E. Chase, "Feedback Control Theory and Arms Races," General Systems 14 (1969), pp. 137-149; Manus I. Midlarsky, "Mathematical Models of Instability and a Theory of Diffusion," International Studies Quarterly 14 (March 1970), pp. 60-84; A.A. Blumberg, "Model for a Two-Adversary Arms Race," Nature 234 (November 19, 1971), p. 158; Chatterjee, "The Equilibrium Theory of Arms Races," op. cit.; Warren R. Phillips, "Where Have All the Theories Gone?" World Politics 26 (January 1974), pp. 155-188; John V. Gillespie, Dina A. Zinnes, and G.S. Tahim, "Foreign Military Assistance and the Armaments Race: A Differential Game Model with Control," Peace Science Society (International) Papers 25 (1975), pp. 35-51; Dina A. Zinnes, John V. Gillespie, and R.M. Rubison, "A Reinterpretation of the Richardson Arms-Race Model," in Mathematical Models in

International Relations, ed. Dina A. Zinnes and John V. Gillespie (New York: Praeger, 1976), pp. 189-217; William Baugh, "Transient-Response Analysis of Richardson Type Arms Race Models," in Mathematical Systems in International Relations Research, ed. John V. Gillespie and Dina A. Zinnes (New York: Praeger, 1977), pp. 221-248; Dina A. Zinnes, John V. Gillespie, and G.S. Tahim, "Modeling a Chimera: Balance of Power Revisited," Journal of Peace Science 3 (Spring 1978), pp. 31-44.

26. Richardson, "Mathematical Psychology of War," op. cit., and also his Arms and Insecurity, op. cit.; Rapoport, "The Mathematics of Arms Races," op. cit.; Paul A. Smoker, "Nation State Escalation and International Integration," Journal of Peace Research 4 (1967), pp. 61-75; and also Smoker's "The Arms Race as an Open and Closed System," Peace Research Society Papers 7 (1967), pp. 41-62; Jeffrey S. Milstein and William C. Mitchell, "Computer Simulation of International Processes: The Vietnam War and the Pre-World War I Naval Race," Peace Research Society Papers 12 (1968), pp. 117-136; D.L. Wagner, R.T. Perkins, and Rein Taagepera, "Fitting the Soviet-U.S. Arms Race Data to the Complete Solution to Richardson's Equations," paper presented to the annual meeting of the Western Regional Peace Science Society, San Diego, February 22, 1974; Michael Olinick, "Stable and Unstable Arms Races," in An Introduction to Mathematical Models in the Social and Life Sciences, ed. Michael Olinick (Reading, MA: Addison-Wesley, 1978), pp. 20-51.

27. Richardson, "Mathematical Psychology of War," op. cit., "Generalized Foreign Politics," op. cit., and Arms and Insecurity, op. cit.; Smoker, "The Arms Race as an Open and Closed System," op. cit.

28. Richardson, "Threats and Insecurity," op. cit., and Arms and Insecurity, op. cit.; Paul Smoker, "A Pilot Study of the Present Arms Race," General Systems 8 (1963), pp. 61-76, and also his "Nation State Escalation and International Integration," op. cit.; and Terry Hopmann and T. Smith, "An Application of a Richardson Process Model: Soviet-American Interactions in the Test Ban Negotiations 1962-63," Journal of Conflict Resolution 21 (December 1977), pp. 701-726.

29. Smoker, "The Arms Race as an Open and Closed System," op. cit.; Zinnes and Gillespie, "Analysis of Arms Race Models," op. cit.; and Rattinger, "Armaments, Detente, and Bureaucracy," op. cit.

30. Jeffrey S. Milstein, "Soviet and American Influences on the Arab-Israeli Arms Race: A Quantitative Analysis," Peace Science Society (International) Papers 15 (1970), pp. 6-27; Jean-Christian Lambelet, "A Dynamic Model of the Arms Race in the Middle East, 1953-1965," General Systems 16 (1971), pp. 145-167; Edward E. Azar, James P. Bennett, and Thomas J. Sloan, "Steps Toward Forecasting International Interactions," Peace Research Society (International) Papers 23 (1970), pp. 27-67; Dieter Ruloff, "Mediation in International Conflicts: A Computer Simulation and Some Results," Peace Science Society (International) Papers 26 (1976), pp. 22-48.

31. Milstein and Mitchell, "Computer Simulation of International Processes," op. cit., and also their "Dynamics of the Vietnam Conflict: A Qualitative Analysis and Predictive Computer Simulation," Peace Research Society Papers 12 (1968), pp. 163-213.

32. Manas Chatterji, "A Model of Resolution of Conflict between India and Pakistan," Peace Science Society (International) Papers 12 (1969), pp. 87-102.

33. Paul Smoker, "A Time Series Analysis of Sino-Indian Relations," Journal of Conflict Resolution 8 (June 1969), pp. 172-191.

34. Barry O'Neil, "The Pattern of Instability among Nations: A Test of Richardson's Theory," General Systems 15 (1970), pp. 175-181.

35. Charles W. Ostrom, Jr., "A Reactive Linkage Model of the U.S. Defense Expenditure Policy Making Process," American Political Science Review 72 (1978), pp. 941-957.

36. Richard A. Brody, Alexandra H. Benham, and Jeffrey S. Milstein, "Hostile International Communication, Arms Production, and Perception of Threat: A Simulation Study," Peace Research Society Papers 7 (1968), pp. 15-40; Martin Patchen, "Models of Cooperation and Conflict: a Critical Review," Journal of Conflict Resolution 14 (September 1970), pp. 389-407; Dagobert L. Brito, "A Dynamic Model of an Armaments Race," International Economic Review 13 (1972), pp. 359-375; Marwin Simaan and Jose B. Cruz, Jr., "A Multistage Game Formulation of Arms Race and Control and Its Relationship to Richardson's Model," in Modeling and Simulation, vol. 4, ed. W.G. Vogt and M.H. Mickle (Pittsburgh: Instrument Society of America, 1973), pp. 149-153; also Simaan and Cruz, "Nash Equilibrium Strategies for the Problem of Armaments Race and Control," Management Science 22 (September 1975), pp. 96-105; as well as Simaan and Cruz, "Formulation of Richardson's Model of Arms Race from a Differential Game Viewpoint," Review of Economic Studies 42 (1975), pp. 67-77; John V. Gillespie and Dina A. Zinnes, "Progressions in Mathematical Models of International Conflict," Synthese 31 (1975), pp. 289-321; Dieter Ruloff, "The Dynamics of Conflict and Cooperation between Nations: A Computer Simulation and Some Results," Journal of Peace Research 12 (1975), pp. 109-121; Michael L. Squires, "Three Models of Arms Races," in Mathematical Models in International Relations, ed. Dina A. Zinnes and John V. Gillespie (New York: Praeger, 1976), pp. 252-273; Dragoslav Siljak, "On the Stability of Arms Races," in Mathematical Systems in International Relations Research, ed. John V. Gillespie and Dina A. Zinnes (New York: Praeger, 1977), pp. 264-296; Richard J. Stoll, "Let the Researcher Beware: The Use of Richardson Equations to Estimate the Parameters of a Dyadic Arms Acquisition Process," American Journal of Political Science 26 (1982), pp. 77-89.

37. Hayward R. Alker, Jr., "The Structure of Social Action in an Arms Race," paper presented to the North American Peace Research Society (International) Meeting, Cambridge, MA, November 10-12, 1968; Jean-Christian Lambelet, "The Anglo-German Dreadnought Race," Peace Science Society (International) Papers 22 (1974); also, Lambelet's "A Numerical Model of the Anglo-German Dreadnought Race," Peace Science Society (International) Papers 24 (1975), pp. 29-48; and his "A Complementary Analysis of the Anglo-German Dreadnought Race, 1905-1916," Peace Science Society (International) Papers 26 (1976), pp. 49-66; and Rein Taagepera, "Stockpile-Budget and Ratio Interaction Models for Arms Races," Peace Science Society (International) Papers 29 (1979-1980), pp. 67-78.

38. Kendall D. Moll, "International Conflict as a Decision System," Journal of Conflict Resolution 18 (December 1974), pp. 555-557.

39. Robert P. Strauss, "An Adaptive Expectations Model of the East-West Arms Race," Peace Research Society Papers 19 (1972), pp. 29-34; Philip Schrodt, "Contagion Effects in Arms Transfers: A Statistical Study," paper presented to the annual meeting of the International Studies Association, Cincinnati, March 25-27, 1982.

40. Taagepera et al., "Soviet-American and Israeli-Arab Arms Races and the Richardson Model," op. cit.; Dina A. Zinnes, John V. Gillespie, and Philip A. Schrodt, "The Arab-Israeli Arms Race: An Empirical Examination," Jerusalem Journal of International Relations 2 (Fall 1976), pp. 28-62; and Taagepera, "Stockpile-Budget and Ratio Interaction Models for Arms Races," op. cit.

41. Michael D. Intriligator, "Strategic Considerations in the Richardson Model of Arms Races," Journal of Political Economy 83 (April 1975), pp. 339-353; Taagepera et al., "Soviet-American and Israeli-Arab Arms Races and the Richardson Model," op. cit.; and Michael D. Ward, "The Differential Paths to Parity: A Study of the Contemporary Arms Race," American Political Science Review 78 (June 1984), pp. 297-317.

42. John A. Ferejohn, "On the Effects of Aid to Nations in Arms Races," in Mathematical Models in International Relations, ed. Dina A. Zinnes and John V. Gillespie (New York: Praeger, 1976), pp. 218-251.

43. Steve Wright, "A Multivariate Time Series Analysis of the Northern Irish Conflict, 1969-1976," Peace Science Society (International) Papers 29 (1979/1980), pp. 28-47.

44. Philip A. Schrodt, "Microcomputers in the Study of Politics: Predicting Wars with the Richardson Arms Race Model," Byte 7 (July 1982), pp. 108-134.

45. J. Despres and P. Dhyrmes, "Defense Budget Interactions: Preliminary Econometric Analysis of U.S., Soviet and Chinese Competition," Confidential, R-1390-PR (Santa Monica, CA: Rand Corp., 1973); Robert Shisko, "Defense Budget Interactions Revisited," Rand Paper P-5882 (Santa Monica, CA: Rand Corp., 1977); John E. Hunter, "Mathematical Models of a Three-Nation Arms Race," Journal of Conflict Resolution 24 (June 1980), pp. 241-252; Alex R. Hybel and Dale B. Robertson, "Assessing the Dynamic Reaction of the USSR to American and Chinese Actions and to Its Nuclear Gap with the USA," International Interactions 4 (1978), pp. 125-154; Richard K. Ashley, The Political Economy of War and Peace: The Sino-Soviet-American Triangle and the Modern Security Problematique (New York and London: Nichols and Pinter Publishers, 1980); Thomas R. Cusack and Michael Don Ward, "Military Spending in the United States, Soviet Union, and the People's Republic of China," Journal of Conflict Resolution 25 (September 1981), pp. 429-469.

46. R. Abelson, "A 'Derivation' of Richardson's Equations," Journal of Conflict Resolution 7 (March 1963), pp. 13-15.

47. Dagobert Brito and Michael Intriligator, "Some Applications of the Maximum Principle to the Problem of an Armaments Race," in Modeling and Simulation, vol. 4, ed. William G. Vogt and Marlin H. Mickle (Pittsburgh: Instrument Society of America, 1973), pp. 140-144; Ferejohn, "On the Effects of Aid to Nations in Arms Races," op. cit.; Hans Rattinger, "From War to War to War: Arms Races in the Middle East," International Studies Quarterly 20 (1976), pp. 501-531; Martin C. McGuire, "A Quantitative Study of the Strategic Arms Race in the Missile Age," Review of

Economics and Statistics 59 (August 1977), pp. 328-339; Shishko, "Defense Budget Interactions Revisited," op. cit.; John V. Gillespie, Dina A. Zinnes, and R. Michael Rubison, "Accumulation in Arms Race Models: A Geometric Lag Perspective," Comparative Political Studies 10 (January 1978), pp. 475-496; Walter Hill, "A Time Lagged Arms Race Model," Journal of Peace Science 3 (Spring 1978), pp. 55-62; Stephen J. Majeski and David L. Jones, "Arms Race Modeling: Causality Analysis and Model Specification," Journal of Conflict Resolution 25 (1981), pp. 259-288.

48. R.L. Hamblin, M. Hout, J.L.L. Miller, and B.L. Pitcher, "Arms Races: A Test of Two Models," American Sociological Review 42 (April 1977), pp. 338-354.

49. Irwin W. Sandberg, "Some Qualitative Properties of Nonlinear Richardson-Type Arms Race Models," in Mathematical Systems in International Relations Research, ed. John V. Gillespie and Dina A. Zinnes (New York: Praeger, 1977), pp. 305-329.

50. Michael D. Wallace and Judy Wilson, "Non-linear Arms Race Models," Journal of Peace Research 15 (1978), pp. 175-192.

51. Charles W. Ostrom, Jr., "Balance of Power and the Maintenance of 'Balance': A Rational Choice Model with Lexical Preferences," in Mathematical Models in International Relations, ed. Dina A. Zinnes and John V. Gillespie (New York: Praeger, 1976), pp. 318-332; Manus I. Midlarsky, "Power and Distance in International Conflict Behavior," in Mathematical Models of International Relations, ed. Dina A. Zinnes and John V. Gillespie (New York: Praeger, 1976), pp. 132-155; Martin W. Sampson III, "Policy Zone: Where Policies Work," in Mathematical Models in International Relations, ed. Dina A. Zinnes and John V. Gillespie (New York: Praeger, 1976), pp. 348-375; Robert H. Kupperman, Harvey A. Smith, Steven C. Goldman, and Mark R. Johnson, "Some Control Theoretic Prospects for International Relations," in Mathematical Systems in International Relations Research, ed. John V. Gillespie and Dina A. Zinnes (New York: Praeger, 1977), pp. 36-48; R. Miller and Stuart Thorson, "Control Concepts in Theories of Governmental Decision-Making: A Structural Approach," in Mathematical Systems in International Relations Research, ed. John V. Gillespie and Dina A. Zinnes (New York: Praeger, 1977), pp. 60-90; Stephen J. Brams, Morton D. Davis, and Philip D. Straffin, Jr., "The Geometry of the Arms Race," International Studies Quarterly 23 (December 1979), pp. 567-588; Dina A. Zinnes, "An Event Model of Conflict Interaction," in Conflict Processes and the Breakdown of International Systems, ed. D.A. Zinnes (Denver: Graduate School of International Studies, 1983), pp. 119-147.

52. George R. Rathjens, Jr. "The Dynamics of the Arms Race," Scientific American 220 (April 1969), pp. 15-25; Edgar M. Bottome, The Balance of Terror: A Guide to the Arms Race (Boston: Beacon Press, 1971); Raymond Hutchings, "Soviet Defense Spending and Soviet External Relations," International Affairs (London) 47 (1971), pp. 518-531; Klause Jurgen Gantzel, "Armament Dynamics in the East-West Conflict: An Arms Race?" Peace Science Society (International) Papers 20 (1973), pp. 1-24; Albert Wohlstetter, "Legends of the Strategic Arms Race, Part I: The Driving Engine," Strategic Review 2 (Fall 1974), pp. 67-92.

53. Smoker, "A Time Series Analysis of Sino-Indian Relations," op. cit.; Norman Z. Alcock and Keith W. Lowe, "The Vietnam War as a Richardson Process," Journal of Peace Research 6 (April 1969), pp. 105-112; Alcock, "Arms Races--A Cause of War?" op. cit., p. 3.

54. Arthur Lee Burns, "A Graphical Approach to Some Problems of the Arms Race," Journal of Conflict Resolution 3 (December 1959), pp. 326-342; Rapoport, "The Mathematics of Arms Races," op. cit.; Boulding, Conflict and Defense: A General Theory, op. cit.

55. Friberg and Jonsson, "A Simple War and Armaments Game," op. cit.; Peter Busch, "Mathematical Models of Arms Races," in What Price Vigilance? The Burdens of National Defense, ed. Bruce M. Russett (New Haven: Yale University Press, 1970), pp. 193-233; Knut Midgaard, "Arms Races, Arms Control, and Disarmament," Cooperation and Conflict 5 (February 1970), pp. 20-25; H. Afheldt and P. Sonntag, "Stability and Deterrence Through Strategic Nuclear Arms," Journal of Peace Research 3 (1973), pp. 245-250; Earl R. Brubaker, "Economic Models of Arms Races: Some Reformulations and Extensions," Journal of Conflict Resolution 17 (June 1973), pp. 187-205; Jean-Christian Lambelet, "Do Arms Races Lead to War?" Journal of Peace Research 12 (1975), pp. 123-128; Brams, Davis, and Straffin, "The Geometry of the Arms Race," op. cit.; Robert G. Muncaster and Dina A. Zinnes, "A Model of Inter-nation Hostility Dynamics and War," Conflict Management and Peace Science 6 (Spring 1982-1983), pp. 19-38.

56. J.G. Dash, "A Condensation in Warpeace Space," General Systems 11 (1966), pp. 143-146.

57. Lewis Fry Richardson, "The Distribution of Wars in Time," Journal of the Royal Statistical Society 107 (1944), pp. 242-250; also of interest in the context of this approach is Richardson's "Statistics of Deadly Quarrels," in Psychological Factors of Peace and War, ed. T.H. Pear (London: Hutchinson, 1950), pp. 237-255; as well as his full-length study, Statistics of Deadly Quarrels, op. cit.; see also Wilkenson's replication of this branch of Richardson's work in Deadly Quarrels: Lewis F. Richardson and the Statistical Study of War, op. cit.

58. Herbert K. Weiss, "Stochastic Models for the Duration and Magnitude of a 'Deadly Quarrel,'" Operations Research 11 (1963), pp. 101-121; William J. Horvath, "A Statistical Model for the Duration of Wars and Strikes," Behavioral Science 13 (January 1968), pp. 18-28; William W. Davis, George T. Duncan, and Randolph M. Siverson, "Stochastic Models of the Distribution of Dyadic War in Time," Technical Report No. 120 (Pittsburgh: Department of Statistics, Carnegie-Mellon University, 1976); and also their "The Dynamics of Warfare: 1816-1965," American Journal of Political Science 22 (November 1978), pp. 772-792; Philip A. Schrodt, "A Mathematical Model of the Persistence of Conflict," International Interactions 8 (1981), pp. 335-348; and also Schrodt's "A Model of Sporadic Conflict," in Conflict Processes and the Breakdown of International Systems, ed. Dina A. Zinnes (Denver: Graduate School of International Studies, 1983), pp. 101-118.

59. Donald H. Kobe, "A Theory of Catalytic War," Journal of Conflict Resolution 6 (June 1962), pp. 125-142; Michael F. Dacey, "A Probability Model for the Rise and Decline of States," Peace Research Society Papers 14 (1970), pp. 147-153; Midlarsky, "Mathematical Models of Instability and a Theory of Diffusion," op. cit.; George T. Duncan and Randolph M. Siverson, "Markov Chain Models for Conflict Analysis: Results from Sino-Indian Relations, 1959-1964," International Studies Quarterly 19 (September 1975), pp. 344-374; Richard Y.P. Li and William R. Thompson, "The 'Coup Contagion' Hypothesis," Journal of Conflict Resolution 19 (1975), pp. 63-88; Philip A. Schrodt, "Richardson's Model as a Markov Process," in Mathematical Models in International Relations, ed. Dina A. Zinnes and John V. Gillespie (New York: Praeger, 1976), pp. 156-175; and also his "Contagion Effects in Arms Transfers: A Statistical Study," paper presented to the annual meeting of the International Studies Association, Cincinnati, March 25-27, 1982; Harvey Starr and Benjamin A. Most, "A Return Journey--Richardson, Frontiers, and Wars in the 1946-1965 Era," Journal of Conflict Resolution 22 (September 1978), pp. 441-467; Benjamin A. Most and Harvey Starr, "Diffusion, Reinforcement, Geopolitics, and the Spread of War," American Political Science Review 74 (December 1980), pp. 932-946; Bruce Bueno de Mesquita, "An Expected Utility Explanation of Conflict Escalation: A Preliminary Analysis," in Conflict Processes and the Breakdown of International Systems, ed. Dina A. Zinnes (Denver: Graduate School of International Studies, 1983), pp. 47-59.

60. William J. Horvath and Caxton C. Foster, "Stochastic Models of War Alliances," Journal of Conflict Resolution 7 (1963), pp. 110-116; Brian L. Job, "Membership in Inter-nation Alliances: 1815-1965: An Exploration Utilizing Mathematical Probability Models," in Mathematical Models in International Relations, ed. Dina A. Zinnes and John V. Gillespie (New York: Praeger, 1976), pp. 74-109; Randolph M. Siverson and George T. Duncan, "Stochastic Models of International Alliance Initiation," in Mathematical Models in International Relations, ed. Dina A. Zinnes and John V. Gillespie (New York: Praeger, 1976), pp. 74-109; Manus I. Midlarsky, "Alliance Behavior and the Approach to World War I: The Use of Bivariate Negative Binomial Distributions," in Conflict Processes and the Breakdown of International Systems, ed. Dina A. Zinnes (Denver: Graduate School of International Studies, 1983), pp. 61-80.

61. Malcolm W. Hoag, "On Stability in Deterrent Races," World Politics 13 (July 1961), pp. 505-527; Robert H. Kupperman and Harvey A. Smith, "Some Computational Techniques and Theorems Concerning Mutual Deterrence Relationships," Journal of Cybernetics 1 (1971), pp. 5-28; and also their "Strategies of Mutual Deterrence," Science 176 (April 7, 1972), pp. 18-23; and, third, their "Deterrent Stability and Strategic Warfare," in Mathematical Systems in International Relations Research, ed. John V. Gillespie and Dina A. Zinnes (New York: Praeger, 1977), pp. 139-166; D. Hunter, "Decision-Making Model in Nuclear Deterrence Theory," Journal of Peace Research 3 (1972), pp. 209-222; William H. Baugh III, "An Operations Analysis Model for the Study of Nuclear Missile System Policies," in Mathematical Models in International Relations, ed. Dina A. Zinnes and John V. Gillespie (New York: Praeger, 1976), pp. 274-305.

62. See, for example, L. Dolansky, "Present State of the Lanchester Theory of Combat," Operations Research 12 (1964), pp. 344-358; or the more recent Martin C. McGuire, "Theoretical Notes on Lanchester Models of Military Tactics, Troop Density, Organization, and Attrition," paper presented to the Conference on Dynamic Models of Conflict, Boulder, CO, October 31 to November 3, 1984.

63. Stephen J. Brams, Morton D. Davis, and Philip D. Straffin, Jr., "The Geometry of the Arms Race," International Studies Quarterly 23 (December 1979), pp. 567-588; Luc Anselin, "A Simulation Framework for Modeling Dynamics in Policy Space," Conflict Management and Peace Science 7 (Fall 1983), pp. 25-38; Claudio Cioffi-Revilla, "Political Reliability in International Relations," in Conflict Processes and the Breakdown of International Systems, ed. Dina A. Zinnes (Denver: Graduate School of International Studies, 1983), pp. 11-45; Piet Rietvold, "Analysis of Conflicts and Compromises, Using Qualitative Data," Conflict Management and Peace Science 7 (Fall 1983), pp. 39-63; Mike Horn, "Arms Races and the Likelihood of War," paper presented to the annual meeting of the International Studies Association, Atlanta, March 27-31, 1984; Stephen J. Majeski, "Expectations and Arms Races," paper presented to the annual meeting of the International Studies Association, Atlanta, March 27-31, 1984.

64. Exceptions can be found, such as the recent work on strategic implications of the strategic defense initiative (SDI) by Stephen J. Brams and D. Marc Kilgour, "Deterrence versus Defense: A Game-Theoretic Model of Star Wars," paper presented to the annual meeting of the American Political Science Association, Chicago, September 3-6, 1987; and also their "Threat Escalation and Crisis Stability: A Game-Theoretic Analysis," American Political Science Review 81 (September 1987), pp. 833-850.

65. Charles Lucier, "Changes in the Values of Arms Race Parameters," Journal of Conflict Resolution 23 (March 1979), pp. 17-39.

66. Miroslav Nincic and Thomas R. Cusack, "The Political Economy of U.S. Military Spending," Journal of Peace Research 16 (1979), pp. 101-115.

67. W. Ladd Hollist and Thomas H. Johnson, "Political-Economic Competition between the United States and the Soviet Union: Three Alternative Simulations," paper presented at the western meeting of the Peace Science Society (International), Stanford, CA, February 26-27, 1979.

68. Alexander Cockburn, The Threat: Inside the Soviet Military Machine (New York: Vintage Books, 1984), pp. 19-21.

69. W. Ladd Hollist, "Alternative Explanation of Competitive Arms Processes: Tests on Four Pairs of Nations," American Journal of Political Science 21 (May 1977), pp. 313-340.

70. Willem Saris and Cees Middendorp, "Arms Races: External Security or Domestic Pressure?" British Journal of Political Science 10 (1980), pp. 121-128.

71. Michael D. Wallace, "Accounting for Superpower Arms Spending," in Threats, Weapons and Foreign Policy, ed. Pat McGowan and Charles W. Kegley, Jr. (Beverly Hills, CA: Sage, 1980), pp. 259-274.

72. Charles W. Ostrom, Jr., "Evaluating Alternative Foreign Policy Decision-Making Models: An Empirical Test between an Arms Race Model and an Organizational Politics Model," Journal of Conflict Resolution 21 (June 1977), pp. 235-265; and also Ostrom's "A Reactive Linkage Model of the U.S. Defense Expenditure Policy Making Process," American Political Science Review 72 (1978), pp. 941-957.

73. Ostrom, "Evaluating Alternative Foreign Policy Decision-Making Models: An Empirical Test between an Arms Race Model and an Organizational Politics Model," op. cit.; and also his "A Reactive Linkage Model of the U.S. Defense Expenditure Policy Making Process," op. cit.; and J.L. Stromberg, "The Internal Mechanisms of Defense Budget Processes--Fiscal 1953-1968" (Santa Monica, CA: Rand Corp., 1970).

74. Stephen J. Majeski, "Mathematical Models of the U.S. Military Expenditure Decision-Making Process" (Ph.D. dissertation, Indiana University, 1981).

75. Dieter Senghaas, Rustung und Militarismus (Frankfurt: Suhrkamp, 1972); and also his "Armament Dynamics and Disarmament," Instant Research on Peace and Violence 6 (1976), pp. 3-17.

76. Gernot Kohler, "Une theorie structuro-dynamique des armements," Etudes internationales (Canada) 7 (March 1976), pp. 25-50; also Kohler's "Exponential Military Growth," Peace Research (Canada) 9 (1977), pp. 165-175; and his "Toward a General Theory of Armaments," Journal of Peace Research 16 (1979), pp. 117-135.

77. Dina A. Zinnes, Barbara J. Hill, David L. Jones, and Stephen J. Majeski, "Modeling Precrisis Interactions," in Missing Elements in Political Inquiry, ed. Judith A. Gillespie and Dina A. Zinnes (Beverly Hills, CA: Sage, 1982), pp. 117-140.

78. Gantzel, "Armament Dynamics in the East-West Conflict: An Arms Race?" op. cit.

79. Raymond Tanter and William C. Potter, "Modelling Alliance Behavior: East-West Conflict Over Berlin," Peace Science Society (International) Papers 20 (1973), pp. 25-41.

80. Graham T. Allison, Essence of Decision: Explaining the Cuban Missile Crisis (Boston: Little, Brown and Company, 1971).

81. Graham T. Allison, "Questions about the Arms Race: Who's Racing Whom? A Bureaucratic Perspective," in American Defense Policy, 4th ed., eds. John E. Endicott and Roy W. Stafford (Baltimore: Johns Hopkins Press, 1977), pp. 424-441; originally published in Contrasting Approaches to Strategic Arms Control, ed. Robert L. Pfaltzgraf, Jr. (Lexington, MA: Lexington Books, 1974).

82. Graham T. Allison and F.A. Morris, "Armaments and Arms Control: Exploring the Determinants of Military Weapons," Daedalus 104 (Summer 1975), pp. 99-129.

83. Rathjens, "The Dynamics of the Arms Race," op. cit.

84. Michael F.H. Dennis, "The Policy Basis of General Purpose Forces," Journal of Conflict Resolution 18 (March 1974), pp. 3-36.

85. Rapoport, "Lewis Fry Richardson's Mathematical Theory of War," op. cit., p. 297.

86. Ibid., pp. 297, 298.

87. Alker, "The Structure of Social Action in an Arms Race," op. cit., pp. 3, 4.

88. Gillespie and Zinnes, "Progressions in Mathematical Models of International Conflict," op. cit.; John V. Gillespie, Dina A. Zinnes, G.S. Tahim, Philip A. Schrodt, and R.M. Rubison, "An Optimal Control Model of Arms Races," American Political Science Review 71 (1977), p. 226.

89. John V. Gillespie, "Optimal Control Theory: A Promising Approach for Future Research," in In Search of Global Patterns, ed. James N. Rosenau (New York: Free Press, 1976), p. 239.

90. C.J. Hitch and R.N. McKean, The Economics of Defense in the Nuclear Age (Cambridge, MA: Harvard University Press, 1963).

91. Michael D. Intriligator, Strategy in a Missile War: Targets and Rates of Fire, Paper No. 10 (Los Angeles: UCLA Security Studies Project, 1967).

92. P.E. Chase, "Control Theory and the Nuclear Arms Race," Bendix Technical Journal (Autumn 1968), pp. 43-54; and Chase's "The Relevance of Arms Race Theory to Arms Control," General Systems 13 (1968), pp. 91-98; as well as his "Feedback Control Theory and Arms Races," General Systems 14 (1969), pp. 137-149.

93. Simaan and Cruz, "A Multistage Game Formulation of Arms Race and Control and Its Relationship to Richardson's Model," op. cit.; also Simaan and Cruz, "Nash Equilibrium Strategies for the Problem of Armaments Race and Control," op. cit.; as well as their "Formulation of Richardson's Model of Arms Race from a Differential Game Viewpoint," op. cit.; and their "Equilibrium Concepts for Arms Race Problems," in Mathematical Systems in International Relations Research, eds. John V. Gillespie and Dina A. Zinnes (New York: Praeger, 1977), pp. 342-366; Brito and Intriligator, "Some Applications of the Maximum Principle to the Problem of an Armaments Race," op. cit.; as well as Brito and Intriligator's "A Game Theoretical Approach to Bureaucratic Behavior," in Dynamic Optimization and Mathematical Economics, ed. P.T. Liu (New York: Plenum Press, 1980), pp. 223-236; Michael D. Intriligator, "Applications of Optimal Control Theory in Economics," Synthese 31 (1975), pp. 271-288; Gillespie and Zinnes, "Progressions in Mathematical Models of International Conflict," op. cit.; John V. Gillespie, Dina A. Zinnes, and G.S. Tahim, "Foreign Military Assistance and the Armaments Race: A Differential Game Model with Control," Peace Science Society (International) Papers 25 (1975), pp. 35-51; Dina A. Zinnes, John V. Gillespie, and Philip A. Schrodt, "An Analysis of a Multistage Game Model of Arms Races," paper presented to the annual meeting of the American Political Science Association, Chicago, August 28 to September 2, 1974; and Zinnes, Gillespie and Schrodt, "The Arab-Israeli Arms Race: An Empirical Examination," op. cit.; Gillespie, Zinnes, Tahim, Schrodt, and Rubison, "An Optimal Control Model of Arms Races," op. cit.; John V. Gillespie, Dina A. Zinnes, G.S. Tahim, Martin W. Sampson III, Philip A. Schrodt, and R.M. Rubison, "Deterrence and Arms Races: An Optimal Control Systems Model," Behavioral Science 24 (July 1979), pp. 250-262; John V. Gillespie, Dina A. Zinnes, Philip A. Schrodt, and G.S. Tahim, "Sensitivity Analysis of an Armaments Race Model," in Threats, Weapons and Foreign Policy, eds. Pat McGowan and Charles W. Kegley, Jr. (Beverly Hills, CA: Sage, 1980), pp. 275-310.

94. Robert H. Kupperman and Harvey A. Smith, "Some Computational Techniques and Theorems Concerning Mutual Deterrence Relationships," Journal of Cybernetics 1 (1971), pp. 5-28.

95. Sampson, "Policy Zone: Where Policies Work," op. cit.

96. Baugh, "Transient-Response Analysis of Richardson Type Arms Race Models," op. cit.

97. Gilmer L. Blankenship, "An Input-Output Model of Human Population Dynamics," in Mathematical Systems in International Relations Research, eds. John V. Gillespie and Dina A. Zinnes (New York: Praeger, 1977), pp. 98-120.

98. Warren R. Phillips, "Command and Control Tools for Crisis Management," in Mathematical Systems in International Relations Research, eds. John V. Gillespie and Dina A. Zinnes (New York: Praeger, 1977), pp. 3-26.

99. James N. Rosenau, "Puzzlement in Foreign Policy," Jerusalem Journal of International Relations 1 (Summer 1976), p. 9.

100. Gillespie and Zinnes, "Progressions in Mathematical Models of International Conflict," op. cit., p. 302.

101. John Von Neuman and Oscar Morgenstern, Theory of Games and Economic Behavior (Princeton, NJ: Princeton University Press, 1944); Anatol Rapoport, Fights, Games and Debates (Ann Arbor: University of Michigan Press, 1961).

102. Rapoport, Fights, Games and Debates, op. cit.; Anatol Rapoport and Albert M. Chummah, Prisoner's Dilemma: A Study in Conflict and Cooperation (Ann Arbor: University of Michigan Press, 1965); Stephen J. Brams, Game Theory and Politics (New York: Free Press, 1975); and also his Superpower Games: Applying Game Theory to Superpower Conflict (New Haven, CT: Yale University Press, 1985).

103. John Von Neuman and Oscar Morgenstern, "Zur Theorie der Gesellschaftspiele," Mathematische Annalen 100 (1928), pp. 295-320.

104. Nash, "The Bargaining Problem," op. cit.

105. Simaan and Cruz, "A Multistage Game Formulation of Arms Race and Control and Its Relationship to Richardson's Model," op. cit., p. 151.

106. Simaan and Cruz, "A Multistage Game Formulation of Arms Race and Control," op. cit.; their "Formulation of Richardson's Model of Arms Race from a Differential Game Viewpoint," op. cit.; and also their "Nash Equilibrium Strategies for the Problem of Armaments Race and Control," op. cit.; Dagobert Brito and Michael D. Intriligator, "A Multiagent Adaptive Approach to Macroeconomic Control," Policy Analysis and Information Systems 2 (1979), pp. 1-9; and their "Nuclear Proliferation and the Armaments Race," Journal of Peace Science 2 (1977), pp. 231-238; and also Dagobert Brito and Michael Intriligator, "A Fixed Point Approach to Multiagent Adaptive Control," Annals of Economic and Social Measurement 6 (1977), pp. 137-145; Gillespie and Zinnes, "Progressions in Mathematical Models of International Conflict," op. cit.

107. Simaan and Cruz, "Formulation of Richardson's Model of Arms Race from a Differential Game Viewpoint," op. cit., p. 74.

108. Nash, "The Bargaining Problem," op. cit.

109. Brian Job, "Dynamic Modeling: Introduction and Synthesis," in Missing Elements in Political Inquiry, eds. Judith A. Gillespie and Dina A. Zinnes (Beverly Hills, CA: Sage, 1982), pp. 89-92, 141-149.

110. James G. March and Herbert A. Simon, Organizations (New York: John Wiley and Sons, 1958); R. Cyert and James G. March, A Behavioral Theory of the Firm (Englewood Cliffs, NJ: Prentice-Hall, 1963); David Braybrooke and Charles E. Lindbloom, A Strategy of Decision (New York: Free Press, 1963); Herbert A. Simon, Models of Man (New York: John Wiley and Sons, 1957); and also Simon's Administrative Behavior: A Study of the Decision-Making Process in the Administrative Organization (New York: Free Press, 1965); John D. Steinbruner, The Cybernetic Theory of Decision: New Dimensions of Political Analysis (Princeton, NJ: Princeton University Press, 1974); Anselm Strauss, Negotiations: Varieties, Contexts, Processes and Social Order (San Francisco: Jossey-Bass Publishers, 1978); Herbert A. Simon, The Sciences of the Artificial (Cambridge, MA: MIT Press, 1969); and also Simon's essay "Artificial Intelligence Research Strategies in the Light of AI Models of Scientific Discovery," Proceedings of the sixth International Joint Conference on Artificial Intelligence, Tokyo, August 20-23, 1979, Vol. 2, pp. 1086-1094; David R. Maines, "Social Organization and Social Structure in Symbolic Interactionist Thought," Annual Review of Sociology 3 (1977), pp. 235-239; Richard O'Toole and Anita W. O'Toole, "Negotiating Interorganizational Orders," The Sociological Quarterly 22 (Winter 1981), pp. 29-41; Herbert A. Simon, "Human Nature in Politics: The Dialogue of Psychology with Political Science," American Political Science Review 79 (June 1985), pp. 293-304.

111. Thomas S. Kuhn, The Structure of Scientific Revolutions, 2d ed. (Chicago: University of Chicago Press, 1970).

112. Rapoport, Fights, Games and Debates, op. cit., p. 85.

113. Thomas L. Saaty, "Mathematical Structures Applicable to Multi-party Conflicts," Peace Research Society Papers 12 (1970), p. 29.

114. Kendall D. Moll and G.M. Luebbert, "Arms Race and Military Expenditure Models," Journal of Conflict Resolution 24 (March 1980), p. 166.

115. Philip A. Schrodt, "The Effects of a Richardson Arms Race on the Distribution of Power in an International System" (Ph.D. dissertation, Indiana University, 1976); Zinnes, Gillespie, and Tahim, "Modeling a Chimera: Balance of Power Revisited," op. cit.; Michael Squires, "Modeling the U.S.-U.S.S.R. Strategic Arms Race" (Ph.D. dissertation, Indiana University, 1982).

116. Robert Huckfeldt, C.W. Kohfeld, and Thomas W. Likens, Dynamic Modeling: An Introduction (Beverly Hills, CA: Sage, 1982), p. 48.

117. Richard J. Stoll, "Two Models of Escalation of Serious Disputes to War, 1816-1976," paper presented to the annual meeting of the International Studies Association, Los Angeles, March 18-22, 1980, p. 79; emphasis in original.

118. Neumann and Morgenstern, "Zur Theorie der Gesellschaftspiele," op. cit.; and see a translated collection of early work by Emile Borel published in Econometrica 21 (1953), pp. 97-117.

119. Richard M. Nixon, The Real War (New York: Warner, 1980).

120. M.S. Sherry, Preparing for the Next War: America Plans for Post-war Defense 1941-1945 (New Haven: Yale University Press, 1977).

121. Perry M. Smith, The Air Force Plans for Peace 1943-1945 (Baltimore: Johns Hopkins Press, 1970).

122. Roger Dingman, "Strategic Planning and the Policy Process: American Plans for War in East Asia, 1945-1950," Naval War College Review 32 (November/December 1979), p. 11.

123. David Rosenberg, "American Atomic Strategy and the Hydrogen Bomb Decision," Journal of American History 66 (June 1979), p. 64.

124. Terry L. Heyns, "Key Issues in National Strategy," in Understanding US Strategy, ed. Terry L. Heyns (Washington, DC: NDU Press, 1983), p. 3.

125. G.A. Miller, E. Galanter, and K.H. Pribram, Plans and the Structure of Behavior (New York: Holt, Rinehart and Winston, 1960), p. 16.

126. Harold D. Lasswell, Politics: Who Gets What, When, How (New York: Peter Smith, 1950); David Easton, The Political System: An Inquiry into the State of Political Science, 2d ed. (New York: Alfred A. Knopf, 1971).

127. James N. Rosenau, The Study of Political Adaptation (London: Frances Pinter, 1981).

128. Noel Kaplowitz, "Psychopolitical Dimensions of International Relations: The Reciprocal Effects of Conflict Strategies," International Studies Quarterly 28 (December 1984), pp. 373-406.

129. Or other network representations, e.g., Michael D. Zisman, "Representation, Specification, and Automation of Office Procedures" (Ph.D. dissertation, University of Pennsylvania, 1977).

130. Earl Sacerdoti, A Structure for Plans and Behavior (New York: Elsevier, 1977).

131. Roger Schank and Kenneth M. Colby, Computer Models of Thought and Language (San Francisco: Freeman, 1973).

132. Jamie G. Carbonell, "POLITICS: Automated Ideological Reasoning," Cognitive Science 2 (1978), pp. 27-51.

133. Richard Herrmann, "The Empirical Challenge of the Cognitive Revolution: A Strategy for Drawing Inferences About Perceptions," International Studies Quarterly 32 (June 1988), pp. 175-203.

134. Thomas Schelling, Arms and Influence (New Haven: Yale University Press, 1966), p. 261.

135. Allison, The Essence of Decision, op. cit.

136. Ibid.; and Glenn H. Snyder and Paul Diesing, Conflict among Nations: Bargaining, Decision Making, and System Structure in International Crises (Princeton, NJ: Princeton University Press, 1974).

137. Steinbruner, The Cybernetic Theory of Decision, op. cit.

138. Nash, "The Bargaining Problem," op. cit.

139. Snyder and Diesing, Conflict among Nations: Bargaining, Decision Making, and System Structure in International Crises, op. cit.

140. Simon, Models of Man, op. cit.; R. Duncan Luce and Howard Raiffa, Games and Decisions: Introduction and Critical Survey (New York: John Wiley and Sons, 1957); William H. Riker and Peter C. Ordeshook, An Introduction to Positive Political Theory (Englewood Cliffs, NJ: Prentice-Hall, 1973).

141. Max Weber, The Theory of Social and Economic Organization, ed. Talcott Parsons, trans. A.M. Henderson and T. Parsons (New York: Oxford University Press, 1947); Simon, Models of Man, op. cit.; March and Simon, Organizations, op. cit.

142. Simon, Administrative Behavior, op. cit.; Allison, Essence of Decision, op. cit.; Cyert and March, A Behavioral Theory of the Firm, op. cit.

143. Weber, The Theory of Social and Economic Organization, op. cit.

144. Simon, Models of Man, op. cit.; March and Simon, Organizations, op. cit.

145. Braybrooke and Lindbloom, A Strategy of Decision, op. cit.

146. Simon, Models of Man, op. cit.; Cyert and March, A Behavioral Theory of the Firm, op. cit.

147. Steinbruner, The Cybernetic Theory of Decision, op. cit.

148. Richard C. Snyder, H.W. Bruck, and B. Sapin, Foreign Policy Decision-Making (New York: Free Press, 1962).

149. Allison, Essence of Decision, op. cit.; Morton Halperin, Bureaucratic Politics and Foreign Policy (Washington, DC: Brookings Institution, 1974).

150. F.J. Roethlisberger and W. Dickson, Management and the Worker (Cambridge, MA: Harvard University Press, 1939).

151. Charles E. Lindbloom, "The Science of Muddling Through," Public Administration Review 19 (Spring 1959), pp. 79-88.

152. Allison, Essence of Decision, op. cit.

153. Simon, The Sciences of the Artificial, op. cit.

154. Strauss, Negotiations: Varieties, Contexts, Processes and Social Order, op. cit.

155. Ostrom, "A Reactive Linkage Model of the U.S. Defense Expenditure Policy Making Process," op. cit.

156. Partha Chatterjee, "Balance of Power in Alliances and Arms Races" (Ph.D. dissertation, University of Rochester, 1971); Schrodt, "The Effects of a Richardson Arms Race on the Distribution of Power in an International System," op. cit.; Ostrom, "Balance of Power and the Maintenance of 'Balance,'" op. cit.

157. Brubaker, "Economic Models of Arms Races," op. cit.

158. Dagobert Brito and Michael D. Intriligator, "SALT and Innovations in Weapons Technology," Public Choice 36 (1981), pp. 41-59.

159. Dagobert Brito and Michael D. Intriligator, "Arms Races: Behavioral and Economic Dimensions," in Missing Elements in Political Inquiry: Logics and Levels of Analysis, eds. Judith A. Gillespie and Dina A. Zinnes (Beverly Hills, CA: Sage, 1982), pp. 93-116.

160. Michael D. Intriligator and Dagobert Brito, "Heuristic Decision Rules, the Dynamics of the Arms Race, and War Initiation," paper presented to the conference on Dynamic Models of Conflict, Boulder, CO, October 31 to November 3, 1984.

161. John P. Crecine and Gregory W. Fischer, "On Resource Allocation in the U.S. Department of Defense," in Political Science Annual: An International Review, ed. C.P. Potter (Indianapolis: Bobbs-Merrill, 1973), vol. 4, pp. 181-236.

162. For example, Strauss, Negotiations: Varieties, Contexts, Processes and Social Order, op. cit.; and Majeski, "Expectations and Arms Races," op. cit.

163. Jean-Christian Lambelet, "Do Arms Races Lead to War?" Journal of Peace Research 12 (1975), pp. 123-128; and also his "Arms Races as Good Things?" paper presented to the Conference on Dynamic Models of Conflict, Boulder, CO, October 31 to November 3, 1984.

164. Roslyn L. Simowitz, "Alliances and the Theory of Collective Goods: An Evaluation," in Mathematical Models in International Relations, eds. Dina A. Zinnes and John V. Gillespie (New York: Praeger, 1976), pp. 333-347.

165. See, for example, Dennis, "The Policy Basis of General Purpose Forces," op. cit.; Ostrom, "Evaluating Alternative Foreign Policy Decision-Making Models," op. cit.; and also Ostrom's "A Reactive Linkage Model of the U.S. Defense Expenditure Policy Making Process," op. cit.

166. James N. McCormick, "Evaluating Models of Crisis Behavior," International Studies Quarterly 19 (March 1975), pp. 17-45.

167. Allison, "Questions about the Arms Race: Who's Racing Whom? A Bureaucratic Perspective," op. cit.; also Allison and Morris, "Armaments and Arms Control: Exploring the Determinants of Military Weapons," op. cit.

168. Charles Lucier, "Changes in the Values of Arms Race Parameters," Journal of Conflict Resolution 23 (March 1979), pp. 17-39.

169. Squires, "Modeling the US-USSR Strategic Arms Race," op. cit.; and his dissertation with a very similar title, "Modeling the U.S.-U.S.S.R. Strategic Arms Race," op. cit.; Mathew D. McCubbins, "Policy Components of Arms Competition," American Journal of Political Science 27 (August 1983), pp. 385-406.

170. Majeski, "Mathematical Models of the U.S. Military Expenditure Decision-Making Process," op. cit.; also his "Dynamic Properties of the U.S. Military Expenditure Decision-Making Process," Conflict Management and Peace Science 7 (Fall 1983), pp. 65-86; and his "Mathematical Models of the U.S. Military Expenditure Decision-Making Process," American Journal of Political Science 27 (August 1983), pp. 485-514.

171. Ibid.

172. Intriligator and Brito, "Heuristic Decision Rules, the Dynamics of the Arms Race, and War Initiation," op. cit.; and also their "Non-Armageddon Solutions to the Arms Race," CISA Reprint No. 1 (Los Angeles: University of California, Center for International Strategic Affairs, 1984); as well as Brito and Intriligator, "SALT and Innovations in Weapons Technology," op. cit.

173. Rapoport, Fights, Games and Debates, op. cit.

174. John Harsanyi, "Mathematical Models for the Genesis of War," World Politics 14 (1962), p. 698.

175. Simon, Models of Man, op. cit.

176. Dina A. Zinnes and John V. Gillespie, "Introduction to Decision-Making Models," in Mathematical Models in International Relations, eds. Dina A. Zinnes and John V. Gillespie (New York: Praeger, 1976), p. 313.

177. Schelling, The Strategy of Conflict, op. cit.; and Harsanyi, "Mathematical Models for the Genesis of War," op. cit.

178. Allison, Essence of Decision, op. cit.

3 METHODS FOR REPRESENTING INTERACTION IN ARMS ACCUMULATION

Chapter Three considers the methods that scholars have used to study armaments accumulation processes. Scholars of the Richardson arms race Tradition have hewn closely to the epistemology of logical positivism, the philosophical basis of modern science. A set of animal parables will help show how a perhaps excessively vigorous interpretation of this epistemology can impede the progress of arms studies. This leads to several conclusions on epistemological and ontological challenges facing arms analysis. From a more technical perspective, dynamic behavior (such as the behavior of armaments accumulation) can be discussed in terms of the "state transformation schema." The traditional data structures and control structures used in arms analysis to represent the state transformation schema impose some unnecessary limitations. This observation sets the stage for a consideration of alternative state spaces, data structures, and control structures that subsume the traditional conceptions. We also will have occasion to consider the techniques of "knowledge engineering," which more easily permit manipulation of these increasingly complex representations. Novel mathematical formalisms are constantly assessed for utility in representing state changes in the empirical universe of international relations. We will explore here some of these new formalisms, namely first- and second-order production systems. In recent years these new techniques have begun to come into use in the study of international conflict, providing the technical foundation necessary to model ideas about strategic planning. This capacity to model planning processes in arms accumulation deserves further attention.

3.1 What Kind of Analogy Are We Dealing with Here?

Three analogies, or parables if you will, tell a story of some differences between the inanimate world of the physical sciences and the animated world of the social sciences. We will consider ants on a beach, cats in pain, and the ritual combat of natives. The world of the living necessarily involves concepts of meaning and purpose--concepts inadmissible to behaviorism and, unsurprisingly, not amenable to representation in the methodology of the physical sciences.

Social science does not deal with a cluster of atoms in a gaseous state or with the energy levels of a quantum. It deals with animate, purposive entities, capable not only of reaction but also action. The "billiard ball model" is not a good one here. To treat social entities as being solely of the same cloth as are objects of the physical world is to miss the purposes infusing social systems. To view the human world as ricochetting sequences of external causal relations is to neglect the dimensions of meaning and purpose inherent in human affairs. Atoms have no meaning, no purpose, and the methods used to treat atoms are not conceptually appropriate to the analysis of meaning and purpose. This conceptual chasm separates explanation from understanding.

3.1.1 Ants on the Beach

Herbert Simon in his 1968 Compton Lectures at MIT told a story about an ant walking across a beach deeply carved by wind and wave.[1] Moving across the rugged terrain, the ant's path becomes impossibly twisted and complex as it jogs and jags back and forth, up and down around obstacles taking advantage of natural pathways. One can imagine a three-dimensional representation of the ant's path through a Euclidean space, but to discover the polynomial equation describing that path is a difficult and thankless task. Yet if one steps back to look at the path of the ant in terms of the context within which it occurred, one sees an ant whose tendency is to head North, toward the nest. It is the complexity of the surface of the beach--the ant's environment--causing the behavior to appear complex. Abstract out the environment and a description of the behavior of the ant--far from the complexity of trying to describe its path through a Euclidean space--is quite simple: to head home. Behavior is thus seen in Simon's perspective as an "artificial" product, resulting from the meeting of purpose and environment. Simon presented this famous hypothesis:

> *An ant, viewed as a behaving system, is quite simple. The apparent complexity of its behavior over time is largely a reflection of the complexity of the environment in which it finds itself.*

The purport of introducing this example here is merely to dramatize the critical importance of the conceptual perspective brought to the study of a given instance of behavior. Under the rubric of the purely positive conceptual framework, one can map behavior in terms of a trajectory through an abstract space; this is perhaps an essential step in the emergence of explanation and understanding from description. But about whether this largely inductive procedure can lead by itself to theoretical understanding, there is much more question.

> Description is of course important. Even more, it is essential to science that we have accurate descriptions available. But there is another ingredient required, one that we seem to forget and rediscover in every generation of psychologists, at least since Brentano's Act first competed with Wundt's Content. Life is more than a thing, an object, a substance that exists. It is also a process that is enacted. We have certain choices in our approach to it.[2]

As some now-unknown economic wag once quipped, "statistics are like bikinis . . . what they reveal is interesting but what they conceal is vital." They conceal, in their seductively veiling assumptions, a lack of potential for structural transformation through purposive adaptation.

3.1.2 Cats in Pain

Peter Winch asks us to compare representations of the reaction of a cat that is seriously hurt and of a tree that is being chopped down.[3] Is the difference in the reactions of the cat and the tree to their travails one of degree or is it a difference in kind? We can describe the tree being chopped down in terms of a set of time-space coordinates. This is a useful thing to do, supposing you value staying out of the path of falling timber. We can also represent the behavior of the cat in a similar fashion, that is, with a set of time-space coordinates describing the complex movements of the wounded cat. But would this make sense in the same ways as it does with the tree? Alternatively, one could describe the behavior of the cat as "writhing." Winch points out that though both of these are ways of describing the cat in pain, the one cannot be substituted for the other. No matter how detailed our mechanical description of the movements of the cat in space-time, the description that includes the "concept of writhing belongs to quite a different framework than that of the concept of movement in terms of space-time coordinates. . . . Anyone who thought that a study of the mechanics of the movement of animate creatures would throw light on the concept of animate life would be the victim of a conceptual misunderstanding."[4]

Cats learn and adapt. Their behavior is molded by their environment in a complex set of relationships. A cat can learn to overcome environmental obstacles impeding the satisfaction of its wants by engaging in new and different behaviors. In this way the cat and other forms displaying individual adaptive learning behavior are qualitatively distinct from trees, rocks, and other forms that "adapt" through only the production of adaptively superior offspring, if at all. Ashby's inexperienced young kitten, like all young kittens fascinated with flame, approaches the fire too closely and gets burned: "Contrast this behavior with that of the experienced cat: on a cold day it approaches the fire to a distance adjusted so that the skin temperature is neither too hot nor too cold. If the fire burns fiercer, the cat will move away until the skin is again warmed to a moderate degree. If the fire burns low the cat will move nearer. . . . Such behavior is 'adapted.'"[5] This is adaptation through "learning," or in Ashby's terminology, "ultra-stability." Ultra-stable systems are able to adapt their rules of behavior to accommodate change in the environment. It was in this same spirit that another thinker once commented that "Mankind is surely more complicated than lifeless matter."[6] The concepts of meaning, intentionality, and purposive adaptation are what make it more complicated.

The mathematical approach chosen by Richardson and his students involves conceptualization of social reality using the framework of physics; this conflicts with the importance of intentionality and purposive adaptation. The variables in the regression equation represent conceptual "black boxes." The regression modeler does not care and is not interested in how the transformation is effected, but rather focuses on the function (F) mapping input (i) onto output (o) in a homomorphic relationship. I --> BB --> O. F(o(i)) does not tell us what happens in order that I --> O; it says only that I --> O, but not why or even how I --> O.

3.1.3 Natives in Ritual Combat

Gregory Bateson journeyed to Oceana to study the anthropology of native cultures. He brought back with him the concept of "schismogenisis." Bateson formally defined this concept in terms of a category he called "progressive differentiation."

> To this category may be referred all those cases in which the individuals in two groups A and B have the same aspirations and same behavior patterns, but are differentiated in the orientation of these patterns.... Thus a position is set up in which the behavior X,Y,Z is the standard reply to X,Y,Z. This position contains elements which may lead to progressive differentiation or schismogenisis along the same lines. If, for example, the patterns X,Y,Z include boasting, we shall see that there is a likelihood, if boasting is the reply to boasting, that each group will drive the other into excessive emphasis of the pattern, a process which if not restrained can only lead to more and more extreme rivalry and ultimately to hostility and the breakdown of the whole system.[7]

The notion of progressive differentiation entered into the language of systems analysis as the concept of "feedback." "Symmetrical schismogenisis" is positive feedback, and "complementary schismogenisis" is negative feedback. Schematically, the action of a linked pair of positive feedback loops forming a system of symmetrical schismogenisis can be expressed in terms of the TOTE (Test, Operate, Test, Exit) representation, as shown in Figure 3.1.[8]

3.1.4 Critique

This is a very simple, economical way to represent behavior. It tells us something important about the relationship between Bateson's two Imatul tribesmen in escalating ritual combat, *viz.*, that the two natives are linked in a single system by a feedback process that creates an effect that neither could achieve in isolation from the other. It does not, however, tell us why they are in combat. Why are they performing this ritual? Does an actual dispute over property or rights lie at the base? Or is it a religious rite connected to hierarchical social structures? It might be a traditional medical procedure, or a political debate. The point, of course, is that the feedback process representation contains no information that would allow a determination on this point. We have no idea, on the basis of knowing that a feedback process exists, what was the etiology of the social ritual, where it came from, or where it might come to rest.

In a trenchant critique of this conceptualization, Hayward Alker, Jr., professes to be "tired of the incredibly reductionist views of social modelers who think a few feedback loops, or non-linear optimal control algorithms are the main or sole worthwhile contributions of the cybernetic tradition to contemporary social science."[9] Alker argues that the "living, reproductive quality, the mutable viability of social relationships" is really the more challenging and interesting issue in the analysis of social systems, a more significant research goal than that of describing the mechanical stability of social structures. While the homeostatic and structural qualities underlying

Figure 3.1
TOTE Representation
of Symmetrical Schismogenisis

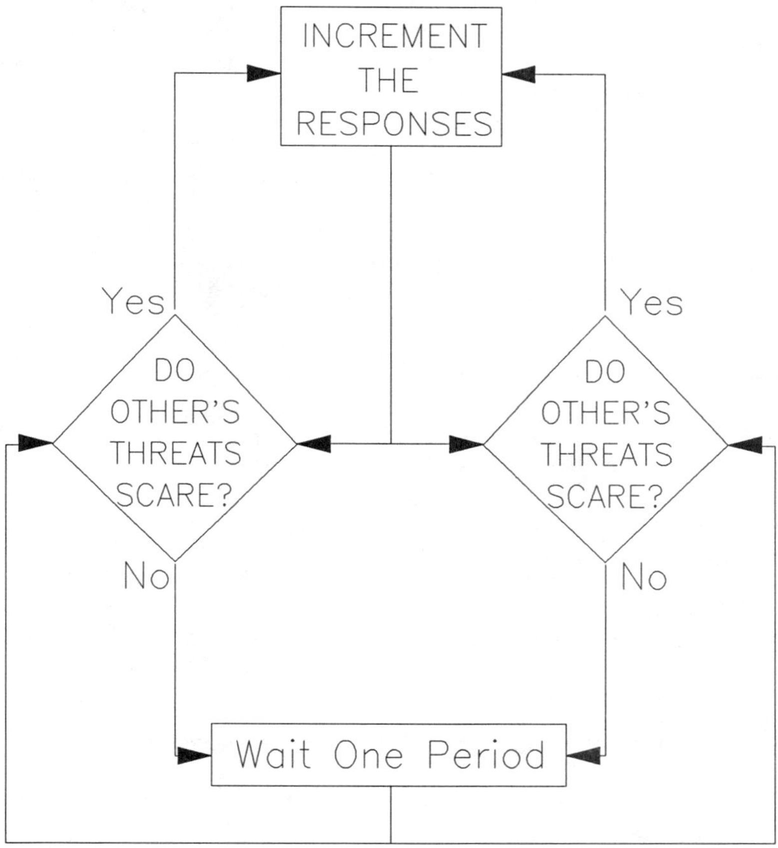

the invariance of ahistorical cross-sections of social behavior may be significant, one might also have a legitimate interest in an understanding of the possibilities for system transformation. It is important to know what we *can* change, as well as what we cannot.

A serious problem with the purely positive interpretations of behavior--such as schismogenisis or feedback--is that these representations do not contain very much information. Social interaction is complex no matter what perspective one adopts, and perspectives relying on highly abstracted segments of the total information set have a tendency to be so elegant and parsimonious that they don't really tell us much about the particular case.

When Bateson discovered Richardson's mathematical models of conflict, he believed that Richardson had captured in his modeling efforts the precise formal relationship between "symmetrical schismogenisis" (the reaction model) and "complementary schismogenisis" (the submission model).[10] The problem again is that the formal representation of social behavior in terms of a simple system of equations based on external covariance relationships does not penetrate those areas most fundamental to human affairs: values, meaning, purpose, plans, and change.

The critics of Aristotilian physics who eliminated the teleological aspects of that theory enabled the modern physical sciences to emerge in all their impressive scope. Over the course of the millennia, however, this has given way to modern critics of teleology whose effect has been to hamper the emergence of the social sciences. Inanimate systems can be profitably modeled without reference to goal-seeking agents, but social systems are suffused with goal-seeking behavior and hence with a teleological aspect. Politics is all about competing goals, and in the study of politics one must consider these subjective teleological aspects. Humans seek ends and goals, and humans create institutions to further their individual and collective goal-seeking behaviors. These behaviors and institutions and the interrelations among them are, or should be, a core preoccupation of political science. We should put teleology back into our social science models.

As Weber argues, "every artifact, such as for example a machine, can be understood only in terms of the meaning which its production and use have had or will have for human action; a meaning which may derive from exceedingly various purposes. Without reference to this meaning such an object remains wholly unintelligible."[11] The task in international relations research, no less than in the other social sciences, is the interpretation of action in terms of its meaning. That this task is challenged by our methods may become clearer after a consideration of some epistemological and ontological aspects of arms accumulation research.

3.2 The Metaphysics of Arms Accumulation Research

Any adequate discussion of the methodology of armaments accumulation research must sooner or later ground itself in metaphysics. An epistemological reconstruction of the Richardson Tradition of armaments accumulation indicates difficulties in the development of a science of arms study, particularly in one key area: emergence of a shared paradigm of arms race behavior. This difficulty is rooted in the problem of intentionality.

3.2.1 Epistemological Reconstruction

Philosopher of science Irme Lakatos has described the growth of knowledge in terms of his "methodology of scientific research programmes."[12] According to this view of espistemology, a scientific enterprise is constituted around a central body of unchallenged assumptions (the "hard core" or "negative heuristic") concerning the phenomenon of interest in that particular area of investigation. This hard core of "unassailable" (or unassailed) premises is surrounded by sets of derivative hypotheses or propositions. The set of derivative hypotheses, called the "protective belt" or "positive heuristic," is subject to testing, falsification, and modification or rejection (hence, the adjective "positive").

The "scientific" character of the enterprise, of the research program, is retained so long as the development of the positive heuristic continues to be progressive, which is to say, as long as the explanatory scope of the theoretical system continues to expand. However, when the protective belt of derivative propositions derived from the unassailable primitive assumptions fails to generate increasing predictive capability, then the research program enters a critical phase, called "degeneration." A principal characteristic of a scientific research program in degeneration is that the negative heuristic of the research program, the hard core of primitive assumptions, will be subject to attack and may be radically altered or rejected altogether.

When a scientific research program is in a degenerative phase, there will be much turmoil in the scientific community bound by the enterprise. Some will view challenges to the negative heuristic as a natural element of the scientific process, while others may tend to resist the undermining of the long-standing, often unspoken assumptions shared by practitioners of the science. Often there will be no consensus as to the state of knowledge in the field and, certainly, at least in the early part of the degenerative phase, no agreement on the fact that a critical period has been entered upon. Until the degenerative crisis is in full bloom, moreover, the precise content and identity of the negative heuristic may remain generally unrecognized. It is during such a critical period in the development of a research program that a "progressive problem shift" may occur. The problem shift refers to a change in some element attaching to the hard core of the research program, and may result in a regeneration of the research program (progressive) or merely continue and/or accelerate the degeneration process (degenerative).

Following this Lakatosian epistemology, we will proceed to reconstruct the history of the Richardson Tradition of armaments analysis. It will become apparent that the tradition has moved through several critical phases of development. The Richardson Tradition faces continued paradigmatic crisis even as the fundamental assumptions of the negative heuristic remain in place.

The core of the negative heuristic in the Richardson Tradition is the Doctrine of Mutuality. In Kuhn's terms, this is the consensual paradigm. For purposes of the present exposition, we will assume that the three tenets of logical positivism as elaborated by Georg Von Wright--methodological monism, the hypothetico-deductive method, and mathematical idealism--form a band of supporting assumptions included in the outer belt of the negative heuristic.[13]

Beyond the negative heuristic has been an evolving collection of hypotheses that constitutes the positive heuristic of the Richardson Tradition. The review of the Richardson literature undertaken in Chapter Two suggests that there has been a distinct process of cumulation, of sorts, in the positive heuristic of the tradition. This forms what might be referred to as a process of "stratified cumulation," piling up in four "layers." In each layer, investigation focused on a particular phase of modeling: the Basic, Autonomous, Mixed, and Teleological phases. Within each layer, there appears to have been a period of "normal science."[14] During the normal period of each layer, the dominant positive heuristic guided puzzle-solving activity until recurrent unsatisfactory research results led to a crisis and a change in the positive heuristic.

The first layer is constituted by the Basic Model and its variants. The normal period for the Basic Model is represented in Richardson's work and in the early work of Smoker. The second layer is constituted of Autonomy Models. Here the normal period can be identified, in temporal terms, as a brief interval in the late 1960s and early 1970s. Investigators quickly realized, however, that models combining features of mutuality and autonomy performed better in many cases. A conscious or unconscious desire to reincorporate elements of mutuality explanations may also explain part of the move from the second layer to the third, because the Autonomy Models represented the response to the first degeneration crisis in the Richardson Tradition. The Autonomy Models directly challenged the hard core of the research program, the Doctrine of Mutuality. Thus, the second layer can also be seen as a period of crisis, a time of transition between the pure mutuality of the Basic Model and the mix of mutuality and autonomy in the combination models.

Yet at least three factors seem to have combined to create a crisis in the third layer of cumulation in the Richardson Tradition. The first was the persistence of questionable empirical performance of the models. The second concerned the continuing preoccupation with the technical problems of estimating parameters in the systems of equations. The third and perhaps most potent factor was continuing suspicion of the lack of any notion of national goals in the third layer models. This led to the emergence of the fourth layer of cumulation in the Richardson Tradition, the Models of Teleological Accumulation.

In theory, the models of the fourth layer should have represented a progressive problem shift. That they did not is dependent upon several factors. The technical difficulties associated with the models of the third layer were not absent from the work of the fourth layer. It is entirely possible to question the degree of empirical confirmation achieved for the models of the fourth layer, as well. The fourth layer succeeded in meeting only the last of three problems plaguing the third layer, that is, the incorporation of some notion of national goals. At the same time, however, the teleological models of the fourth layer unwittingly introduced the seeds of a continuing crisis of degeneration in the research program of the Richardson Tradition. The seed is teleonomy.

It will be recalled that one element of the hard core of the Richardson Tradition is the trilogy of tenets defining logical positivism. The introduction of teleology seems to directly challenge at least one central tenet: methodological monism. Ideas of teleology imply a move into the non-phenomenological world of intentionality, of goals and plans and purposes. This is uncongenial to logical positivism, contributing to the continuing degeneration crisis of the Richardson Tradition. Until this challenge is dealt with, the crisis of degeneration is likely to continue the threaten the hard core of the Richardson Tradition, the Doctrine of Mutuality.

There remains much controversy on the epistemological status of the Richardson Tradition. Criticizing Kuhn's term, "paradigm,"[15] Alker set forth the notion of a "research paradigm complex" in an attempt to identify the epistemological status of the Richardson Tradition.[16] Ashley refers to Alker's criteria in challenging his assessment.[17] Ashley argues that the Richardson Tradition is not a paradigm because (1) the literature is not cumulative in a linear sense; (2) it is doubtful that the "scholarly community" of the Richardson Tradition is bounded and bonded to the degree implied by Alker's criteria; and (3) that genuine crises are not possible because puzzles can't be distinguished from anomalies due to the lack of adherence to and specificity of metascientific beliefs, exemplars, positive heuristics, and symbolic generalizations. He settles on the term "pre-paradigmatic research nucleus."

The foregoing literature review would not seem to support the first two of Ashley's three points. First, it has been shown that there has been a process of cumulation in the literature of the Richardson Tradition. Second, the several hundred investigators reviewed exhibited a remarkable uniformity of methodology and conceptual perspective, apparently indicating a community. On the third point, it is an open question whether workers in the Richardson Tradition will be able to distinguish between puzzles and anomalies and hence respond appropriately to continuing crisis within the discipline.

Teleonomic or teleological models are indicated by theoretical and empirical results, but this conflicts with fundamental elements of the negative heuristic, namely, methodological monism. The Richardson research program seems to be responding to this continuing crisis with a progressive problem shift requiring the elaboration of more sophisticated teleonomic models. Models in this fifth layer of cumulation in the Richardson Tradition are likely to increasingly involve approaches based upon finite state machines.

3.2.2 Methodological Monism

A fundamental tenet of the logical positivist epistemology is the idea of methodological monism, the belief that the experimental methods of the physical sciences are precisely appropriate for the study of all scientific problems, including those of the social sciences.[18] Methodological monism is tied to the notion that the substance of the physical world is exactly the same as the substance of the social world. Many behavioral social scientists, with an extreme phenomenological bias, would with Ludwig Wittgenstein deny the relevance of anything to which an electrode can not be attached. It may well be true (as Albert Einstein argued in support of the position that there exists a physical mechanism governing changes in the energy levels of atomic particles) that God does not shoot dice with the Universe. Einstein failed to add that many humans do shoot dice, and some seem to rather enjoy the game.

The Richardson model of arms races captures the mechanistic logic of two nations locked into a conflict spiral. It is a representation of a deterministic phenomenon, just as Einstein's quantum energy transitions are deterministic. There is a mechanism. In a machine, there is no chance, no choice.

What precisely do we mean when we say that arms race models to date have been "mechanical"? My dictionary defines mechanical as (1) of or relating to machinery or tools; (2) produced or operated by a machine. A later, more elaborate definition is more revealing: (4b) relating to the quantitative relations of force and matter; (5) caused by, resulting from, or relating to a process that involves purely physical change. "Purely physical change" is a phrase to which attention is drawn.

Surveys of arms race modeling methodologies over the last half century demonstrate that the mathematical structures used to date, whether discrete (e.g., difference equations, combinatorics, stochastic processes, networks, graphs, and linear algebra) or continuous (e.g., geometry, the calculus, topology, control theory, and dynamic optimization) representations, virtually all originated with and were created to solve the problems of the physical sciences.[19] The same might be said of fuzzy sets, catastrophe theory, turbulence and strange attractors, and several other advanced mathematical structures and concepts that, though not yet carefully explored, are likely to be applied to arms race theory by enterprising investigators in the near future.

3.2.3 Defining the Challenge

Values and goals--the substance of social and political problems--are not physical entities. Increasingly, social scientists recognize that these things are not particularly amenable to representation and manipulation by physicalistic methods, though they are crucial to the process of explanation and understanding in the social sciences. The social world is not made up of only passive particles, but by active ones as well. It is the "active social particle," so to speak, that accounts for the purposive nature of social life, for the mutability of social relations and for the adaptation of social forms. The mutability of social relations is something ill suited to analysis by physics, which after all was designed not to discover change but rather invariance.

The challenge to the social sciences lies in finding and applying a "home-grown" genre of mathematical structures. These should be structures that are specifically designed to work with the kinds of data and kinds of problems facing the social sciences. Game theory, notwithstanding its limitations, is a premier exemplar of this imperative. Techniques for representing and manipulating qualitative data exist and are proliferating rapidly. It is safe to assume that they will eventually be appropriated by our discipline, and in fact this is happening already.

Looking back on the development of mathematics over the past several centuries, one can see that mathematical structures have been developed hand in hand with pure and applied research endeavors in a variety of scientific problem domains, especially the "queen" of the physical sciences, theoretical physics. From the history of the development of the social sciences over the course of the past two centuries or so, it seems likely that the same process is occurring here as well, if only a little more gradually. This may be taken as an indication that the subject matter of the social sciences is somewhat more difficult than that of the physical world. Yet this does not necessarily constitute cause to abandon the positivist tenet of mathematical idealism, if only we can be a bit more inventive. Perhaps this requires being more playful with such a young and aspiring science and willing to take risks. As Cioffi-Revilla suggests, "games, fuzzy sets, catastrophes and many developments in some of the other classical structures have come from the basic need to develop formal languages capable of capturing the nuance of empirical universes such as" in international relations.[20]

3.3 Ontological Aspects of Arms Accumulation Research

It is incumbent upon the researcher to make the philosophical underpinnings of his or her work as plain as possible. The ontological philosophy of Georg Wilhelm Von Leibnitz informs the present study.[21] How this shapes the conceptual perspective taken in the present work will be seen directly by considering the following.

Conceptually, the state apparatus in general and military organizations in particular are viewed as being monadic in character, differing from the Leibnitzian monad only in that they are dynamic and adaptive, rather than eternal and unchanging as in the original formulation. These adaptive military monads can be considered in terms of their "structure." Monads have surface structures and deep structures. Surface structures are *artifacts*--they lie at the interface between the organization and the environment. Surface structures are manifest in terms of "military force postures" and the formal organization of the military bureaucracy. Deep structures are purposive or intentional, residing within the organization and defining its nature. Deep structures are described in terms of strategy, plans, and goals. The deep structure of Monad A is conceived to dominate change its own surface structure, while change in its deep structure results from change in the surface structure of Monads B through N, etc.

It may be helpful to restate the Leibnitzian metaphysics in more contemporary terms. Happily, this is easily accomplished by reference to a widely known scientist and one of his most popular works.[22] In his *The Sciences of the Artificial*, Herbert Simon advances an ontology that is, from our perspective, identical to our modified Leibnitizian stance. For Simon, behavior is "artificial," or "artiface," i.e., the product of the meeting of purpose and environment. Here behavior is guided by goals and shaped by contingency. The "inner environment" of intention meets the "outer environment" of the universe, forming a mold at the juncture, the realm of the artificial. Our surface structures, the visible forms of bureaucracy, grow into the mold.

Translated into Leibnitzian terminology, Simon's artificial is equivalent to the surface reflections or surface structure. Leibnitz' active principle is Simon's inner environment, here interpreted in terms of the goals or plans lying at the center of the monad. Finally, the Leibnitzian universe corresponds to Simon's outer environment. This view is contrasted with the traditional positive perspective, which views interaction as an external relation among undifferentiated formal entities, as shown in Figure 3.2. The internal relation among individuated entities in the Leibnitzian cosmology is represented in Figure 3.3.

3.3.1 Machines and the Study of Arms Accumulation

Dynamic systems--such as the international system--are typically discussed in terms of models composed of data structures and control structures. This sections treats the representation of these systems. The matter of how to represent a dynamic system, that is, a system that changes over time, can be presented as a question of "state transformation." The state of any system at a particular point in time is given by the

Figure 3.2

Representation of Positive Cosmology

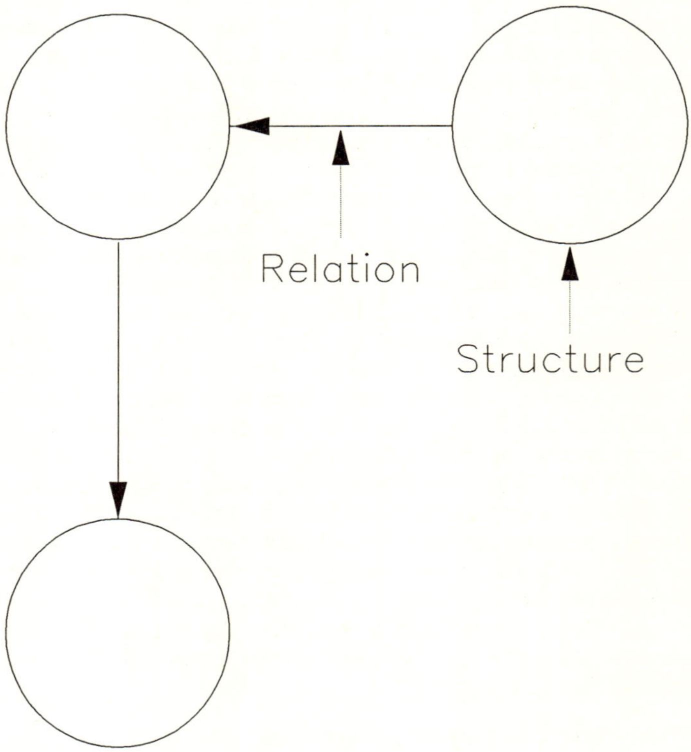

Figure 3.3
Representation of Monadic Cosmology

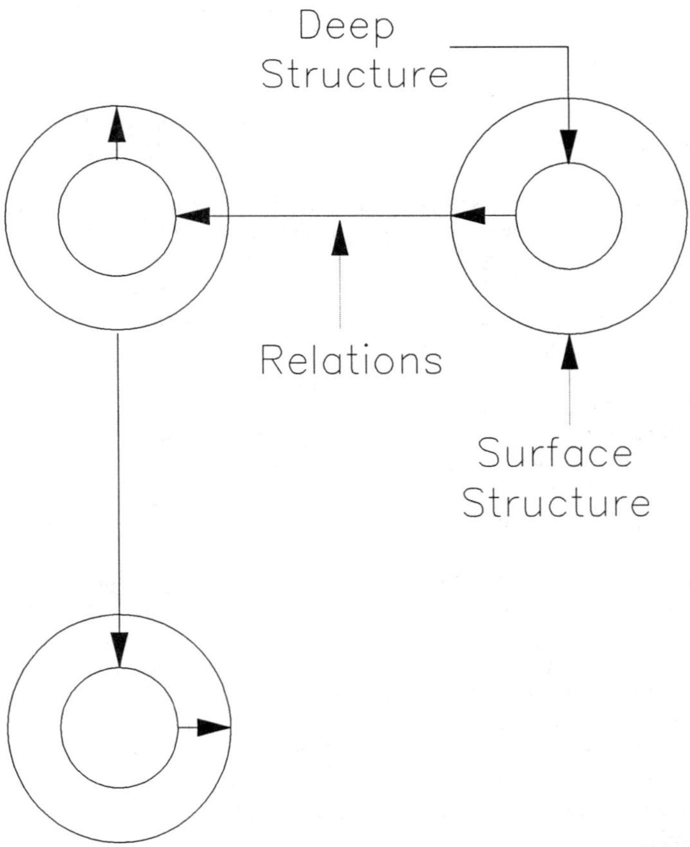

values of the "critical variables" of the system. The entities of interest are the beginning state, the ending state, and what comes in between. As depicted in Figure 3.4, we have an "Initial State" (S_0), a "Goal State" (S_n), and an "Operator" or series of Operators that accomplish the transformation between the initial state and the goal state.

This generic representation of an arbitrary process of state change makes it evident that there is no *a priori* reason why one mathematical structure should be more appropriate than another to the task of realizing the representation. Conversely, some formalisms, by dint of their particular syntax, may be more suited to the task of representing a given problem domain.

3.3.2 Data Structures in the Study of Arms Accumulation

Turning now to data structures, what is the nature of the data structures employed in the study of armaments accumulation? The predominant approach is a bi-variate information set, technically a "2-tuple." That is, there are two pieces of data upon which attention is focused: the dependent variables of annual arms expenditures by each nation. These two pieces of data are real numbers, arranged as scalar vectors on time. Their data *type*, then, is scalar. The general problem is to find equations that describe the covariation of two scalar vectors. The vectors express trajectories through an abstract two-dimensional space, sometimes referred to as the "warpeace space." The typical Richardson model, halted at any point during execution, has the state description (x_t, y_t), which gives the condition or the state of the system at time t. The "x" and "y," of course, represent the nations' arms expenditures.

Figure 2.2 (see Chapter Two, p. 30) describes the state space of a typical Richardson system. This state space or warpeace space is the set of all possible states of the conflict system. One wants to describe the "Initial State," (say, war), a transformational operator (say, a reaction process), and a "Goal State" (say, peace). For example, one might want to describe a system that is initiated from the point (X_t, Y_t)--representing a level of armaments in a state of war--and transformed over time by a reaction process that reaches at time (t') the point ($X_{t'}$, $Y_{t'}$), a point consistent with the state of peace. The relationship accounting for the trajectory traced by Line L in Figure 2.2 is described by the Richardson model (e.g., Equations 1.1 and 1.2 or 2.5 and 2.6).

The first observation here is that this state space is information poor. That is, the vocabulary to describe this system is limited to the simple 2-tuple (x_t, y_t). All we are able to say about a particular system is what value x and y have at time t, where x and y are simply scalars. The scalar data type, it should be noted, is the most elementary type of data structure. This data type is appropriate for representing atomic information, the constituent elements of an atomic universe. Reasons for choosing atomic data (such as dollar expenditures) in arms modeling include the facts that it is there, in a convenient time-series form, and the simple data type is amenable to manipulation with simple types of transformational operators (control structures) based on addition.

Figure 3.4 State Transformation Schema

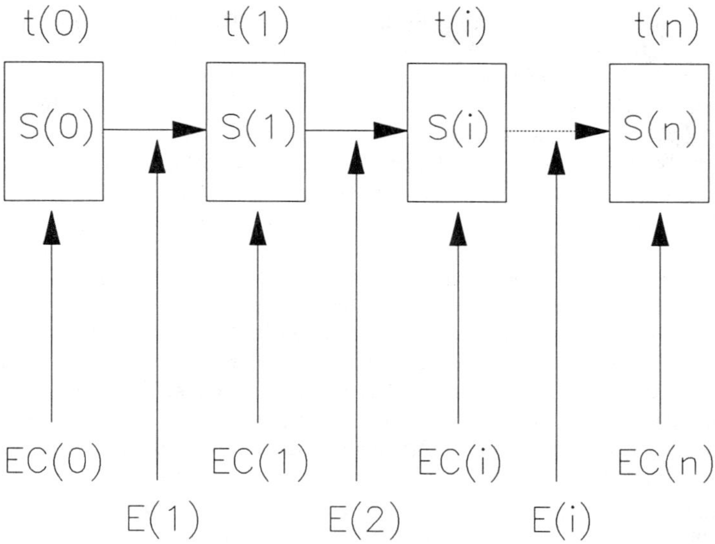

t(n) — time
S(n) — state variables ("surface structure")
EC(n) — enabling conditions ("deep structure")
E(i) — events or operators (transform functions)

In an empirical universe characterized by a situation of armaments accumulation, however, there exists additional data relevant to the questions at hand. This includes some things that are difficult to stick into a scalar vector: the structure of bureaucratic organization, complex collections of partially contradictory goals and motives, webs of plans and counter-plans, dynamic coalitions of overlapping interests, and the psychological dimensions of deterrence. Those data that do not fit into the scalar data structures, however, are represented poorly or not at all in the traditional formalisms. The traditional view is that guns "are seldom fired in anger, tanks and planes do not meet in combat, and there is no general way to express the quantity of defense readiness except in terms of what it costs."[23] The traditional view can no longer be considered sufficient.

After all, as Brito and Intriligator point out, the "interaction between weapons decisions and strategic choices is a fundamental element of the arms race. The strategic options open to a nation depend, to a great extent, on its weapons systems, while conversely, the weapons systems a nation procures depends largely on its strategic goals."[24] Stoll reminds us that "wars are fought by soldiers with weapons, not by accountants with dollars."[25] Nonetheless, with a handful of exceptions to prove the rule, it has been standard procedure in the Richardson Tradition to employ monetary arms expenditure data. This represents a basic and serious challenge to students of the Richardson Tradition. It is a challenge of the first order for the study of arms accumulation to begin using data structures that allow the representation of more complex operational concepts than that of the budget.

A variety of data types are used in data processing and data analysis, such as the scalar, the array, and the digraph.[26] Of these various data types, the scalar is the most elementary, atomic type of structure. Such a structure, again, is appropriate for representing discrete, atomic kinds of things (such as budgets). Such a methodological approach makes good sense when the universe is viewed as a collection of atoms. Scalars are not necessarily appropriate for representing other kinds of things, for example, a strategy. Nor are scalars necessarily the most appropriate structure when one begins to think of the universe in terms of complex relationships and interlinked systems comprising a whole. Here, more complex data structures are required, such as a set of lists linked by pointers. Until the Richardson Tradition succeeds in appropriating these more complex data structures, we will remain unable to capture all the empirical phenomena of interest in this problem domain.

3.3.3 Refining the Challenge

We need a way to deal with two kinds of qualitative data: (1) the murky, mysterious, and mystical world of political purposes, intentions, goals, and plans known as strategy and (2) configurations of military capability known as "force postures." Strategy and force posture are difficult to represent; as the source of system dynamics, however, the relationship among the strategies and force postures of nations in a conflict system stands in need of our analytical attention, regardless of the difficulty of the challenge.

How many different strategies are possible? Depending upon how carefully one cares to examine a particular case, one can elaborate any number of individual military strategies of conflict. But for practical purposes, one may define a relatively small set of military strategies, say five, oversimplifying for the purposes of exposition. These five fairly well cover the gamut of things one can expect to see in politico-military relations among states. Let them be Battle, Control, Defend, Invade, and Power.

Likewise, there exists an infinite number of fine gradations in magnitude and types and other characteristics of force postures. Again generalizing, it is possible to delimit the concept fairly tightly. If one can have a navy, army, and air force in the world under examination, these produce a partial ordering on the three services according to the relative emphasis given to each. This yields a small set of (N!/N-1) or six force postures:

(1) ARMY < NAVY < AIR
(2) ARMY < AIR < NAVY
(3) NAVY < AIR < ARMY
(4) NAVY < ARMY < AIR
(5) AIR < ARMY < NAVY
(6) AIR < NAVY < ARMY

Combining the six force postures and the five strategies produces the matrix shown as Figure 3.5. The matrix describes the state space for any particular actor in the system. There are thirty distinct qualitative state descriptions, and, within each distinct state, there exists an infinite number of possible quantitative gradations, accounting for all possible aggregate levels and compositions of armaments. Thus, such a state space would encompass the traditional bi-variate data structures and then go far beyond by encompassing qualitative information about strategy and orders of battle.

Figure 3.6 describes the state space for a two-actor system composed of the more complex information set described above. Any state description of the system can still be given as (x_t, y_t), where x and y now represent not scalar values for aggregate arms expenditures but rather literal values for strategy and force posture (B3, I4). Under the current definitions, this state space then contains 30^2 or 900 distinct potential qualitative states, with an infinite number of possible quantitative gradations. All of the 900 states may be possible, though some are more likely than others. Some in fact one might suspect to be so very much more likely than others, with a given set of initial conditions, it might be expected that some "natural" patterns would begin to emerge from the data.

The sort of analysis this kind of conceptualization will allow differs substantially from traditional approaches to the problem of armaments accumulation. For example, one can begin to imagine and search for "natural histories" in the patterns of transformation between one strategy and another. One idea might be to apply a type of structure similar to the case grammar approach explored by Dwain Mefford, James Bennett, and others to the kinds strategy transformations put forward by Noel Kaplowitz.[27] This type of conceptualization admits state spaces, and approaches to their representation, which allow study of data structures necessary to conceptually adequate analyses of armaments accumulation processes.

Figure 3.5 Qualitative State Space for a Single-Actor Representation

	1	2	3	4	5	6
BATTLE	B1	B2	B3	B4	B5	B6
CONTROL	C1	C2	C3	C4	C5	C6
DEFEND	D1	D2	D3	D4	D5	D6
INVADE	I1	I2	I3	I4	I5	I6
POWER	P1	P2	P3	P4	P5	P6

Where: BATTLE . . . POWER represent various military strategies, and 1 . . . 6 represent a partial ordering of force postures on three military services.

Figure 3.6 Qualitative State Space for a Two-Actor Representation

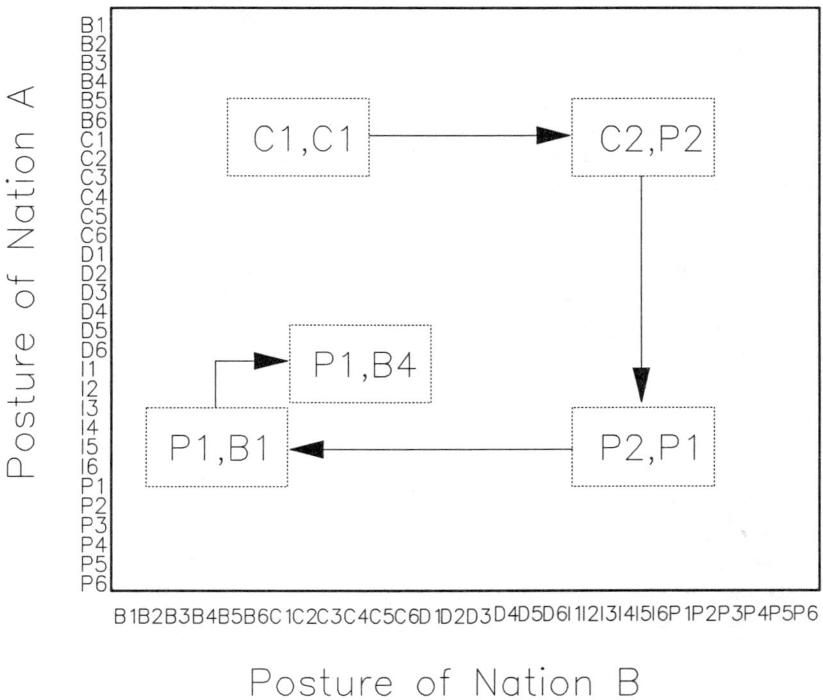

3.3.4 Control Structures of Arms Accumulation

Having described the nature of data structures in the modeling of armaments accumulation, it is now appropriate to consider the control structures governing change in the systems over time. Recall Figure 3.4, a schematic representation of the state transformation process. One notices that the state space includes the Initial State, the Goal State, and any intermediate states between the two. Now the problem is to describe a "thing" that controls the transformation of the system from the initial to the goal state. This thing is called a "control structure." It is comprised of "operators" that operate upon and thus transform the state of the system. The concept is not as remote as it sounds, for in quantitative mathematical systems of the sort commonly used in armaments analysis, one has to consider only two operators: addition and subtraction. These two operators (plus a few variations such as multiplication and division) provide a concise and powerful vocabulary to describe mechanisms for state transformation.

They also impose some severe limitations in the sorts of transformations and states that can be described. What happens if one of the transformations in the system cannot be expressed as an additive relationship? As a trivial example, consider the change of state in the international system on the morning of December 7, 1941. The levels of armaments possessed by Japan and the United States did not change appreciably at the instant the first Japanese warplane attacked Pearl Harbor, but the state of the international system was profoundly changed in a way that is difficult to express in terms of a number.

This is the problem of finding an appropriate metric in which to express the kinds of concepts needed in the study of armaments accumulation processes. How do we represent things like strategy? As summarized by Schrodt, "by and large, the empirical environment of the international system is not one for which social sciences statistics were designed: we can't sample, we can't experiment, we've got too many independent variables, and we have very non-Normal distributions for the stochastic components of our models."[28] It is worth noting that one of the more frequently cited reasons given by policy makers and traditionalist scholars for rejecting the findings of armaments accumulation studies is that the simple mathematical structures cannot capture the richness, subtlety, and complexity of the decision-making process that generates armaments accumulation processes.

There are several approaches to dealing with this deficiency. All of them involve recognizing the value of modern mathematics. As noted, the Richardson Tradition has relied predominantly upon techniques of classical algebraic analysis and the differential and infinitesimal calculi. These mathematical techniques emerged in the seventeenth and eighteenth centuries. Mathematics as a science has undergone several fundamental transformations in the course of developments since that time, most notably with the emergence and consolidation of the theory of groups or sets in the previous and present centuries.[29]

It is fair to wonder why the work of Richardson Tradition has been conducted in a mathematical currency three and four hundred years old. It is natural in this regard to point to the profound influence of the founder of the tradition, who began his labor in a time when the outlines of modern algebra were only just beginning to become clear. This does not, however, excuse a continuing failure to exploit the developments in mathematics and information sciences that have become available over the previous couple of hundred years.

In an original and important contribution, R. Miller and Stuart Thorson deal with precisely this issue.[30] Miller and Thorson examine the abstract structure of control theory and, in what amounts to a critique of applications of optimal control theory in the Richardson Tradition, point out that the student of international politics is primarily concerned with a substantive modeling problem, "not the methodology of solving control problems." They proceed to show that the class of problems solved by optimal control theory is only a small subset of control theory in general. Miller and Thorson are appealing for an update in our mathematical techniques, moving from control structures based on seventeenth- and eighteenth-century mathematics to those of the nineteenth and twentieth. They argue, the "political scientist should not, therefore, be limited to the established detailed models of control theory when addressing the control-related problems of governmental decision making."[31] They suggest one such innovative control structure, the production system. In later work, Thorson and collaborators have extended this idea.[32] In a later section of the present study, precisely this control structure will be applied to a simulation of the role of strategy in arms accumulation.

This discussion of mathematical languages has necessarily been highly abstract. The proposition can be made more concrete by means of two illustrations. Employing either quantitative algebraic analytic languages, or qualitative list-oriented languages, it is a relatively straightforward matter to represent the concept of equality. In the algebraic representation of the classic Aristotilian concept of equality, it is

$$A = A \qquad [3.1]$$

or in a qualitative representation, say a list-oriented syntax,

(EQUAL (A) (A)) [3.2]

What if we want to represent relations between A and A' that are based on concepts other than equality? In classical algebra, our options are severely circumscribed. For example, even as elemental a concept as set membership, difficult to represent and manipulate in quantitative terms, has natural representations in a qualitative approach. Given that A is an apple and B is fruits, it is not accurate to say that A = B. Using qualitative formalisms, however, knowledge engineers have learned how to represent and manipulate the idea of "A IS-A B."[33]

To bring the discussion to a more concrete level yet, one can consider ways to represent an organizational process model. Following Ostrom,[34] one can represent in the traditional algebraic formalism a typical organizational process model composed of current expenditures (Yi^{t+1}, the dependent variable), previous expenditures (Yi^{t}), times a coefficient (Aij) and a random component (Ui^{t+1}, stochastic disturbance).

$$Yi^{t+1} = Aij * Yi^{t} + Ui^{t+1} \qquad [3.3]$$

Although the example is trivial and oversimplified for the purposes of clarity in exposition, this traditional algebraic formalism of the organizational process model might just as easily be rendered in qualitative terms as

$$(\text{EQUAL } Yi^{t+1} (\text{PLUS } (\text{TIMES } Aij (Yi^{t})) Ui^{t+1}) \qquad [3.4]$$

This serves to demonstrate that in many cases a relationship of state transformation operators can be rendered in many different languages. Some may be more efficient representations than others, however, and some ideas may not be expressible in a given language. For example, one might want to define Equation 3.4 as "SCENARIO27," and then incorporate the scenario as an element in a set of scenarios that are used as invoked in a decision-making procedure. Then, say, define a rule as:

WHEN CONDITION (STRATEGIC SURPRISE) IS NOT TRUE,
SET PLANNING_PROJECTION_PARAMETER TO SCENARIO27 [3.5]

Simple enough as a list structure, this would be an extremely cumbersome problem in classical algebraic structures, if it is indeed possible in such a formalism. This example is intended to introduce, in a simple albeit provocative fashion, some of the potential for exploiting the formalisms that have not been extensively applied to social science research in general or international relations research in particular. To understand this new class of formalisms and their potential applications to the problem of interaction in armaments accumulation, it will be helpful to consider the origins of this new methodology. That's what's next.

3.4 AI Techniques and Arms Accumulation Modeling

Perspectives, insights, and techniques developed in artificial intelligence (AI) research are of interest for their potential applications in social science in general and peace science in particular. Though no panacea, some of these techniques seem to be "natural" for application to the problems at hand. One must exercise considerable caution in considering the potential of these techniques, for there is a well-demonstrated tendency to overglorify the actual performance of existing AI technology. This can be seen by examining the history of AI research.

3.4.1 What Is AI?

Partially because it is a very young field of inquiry, AI is as difficult to define as is intelligence itself. AI is centrally concerned with the derivation of structure (or meaning) from unstructured (or unspecified) input. One might characterize the primary transformations currently being studied in AI as: (A) verbal and optical to text (artificial speech and vision), (B) text to meaning (artificial understanding), and (C) meaning to plans (artificial behavior). The "neatness" of these abstract categories, however, belies the rich ferment of contemporary AI research.

Scores of different research foci make such a casual characterization of the AI field perilous. The *Proceedings* of the Sixth International Joint Conference on Artificial Intelligence (IJCAI) in 1979, for instance, list panels and papers under thirty-two separate headings, of which about 90% fit into the broad categories of artificial speech, vision, understanding, and behavior. Another observer abstracts natural language, vision, and knowledge representation as the central foci of AI, accounting for 60% of the Fifth IJCAI *Proceedings*.[35] This diversity within AI research reflects in part the field's complex intellectual lineages, including roots to early work in self-organizing systems, control theory, automata, cybernetics, general systems theory, neuro-physiology, information theory, linguistics, semantics, and cognitive psychology, among others.

Yet it is from within this rich heritage that the ethos of AI emerges. To the extent that there exists a central motif in contemporary AI, the exemplary symbol cluster is purposive behavior. How is the behavior of teleonomic agents shaped by impinging externalities? How are these externalities given meaning in the context of purposeful action? How does meaning relate to the spinning out of adaptive solutions to environmental change? Problem-solving units must be capable of interpreting their stimuli and shaping their responses so as to maintain the essential parameters of their existence against an environment that constantly threatens unmanageable inundations of ambiguous stimuli and combinatorial explosions of possible solutions.

Much of the current work in AI research is related to and/or based upon what is sometimes referred to, somewhat misleadingly, as "natural language processing," more properly known as "list processing" languages. This is a family of computer languages able to perform tasks requiring certain symbolic reasoning capabilities. LISP (from list processing) is the most ubiquitous of these natural or logical programming languages.[36] LISP, PROLOG, and other AI languages involve the manipulation of symbolic strings or qualities, much as BASIC, PL-I, COBOL, and other conventional programming languages involve the manipulation of numeric strings or quantities. Number systems, of course, constitute a category of symbols. Symbols being more general than numbers, the former subsumes the latter. This point is fundamental to an understanding of the relative power and scope of qualitative and quantitative symbol systems.

Two of LISP's most useful characteristics are its generality and flexibility. That is, LISP is a flexible knowledge representation medium because the logic of program flow, within the confines of user-defined and system-defined functions, is largely determined by the purposes of the programmer and the structure of his or her data. In "data-driven" programming, the distinction between the program and the data--so central in normal, quantitative programming--is completely lost. Rather than the program structure as in quantitative programming, it is the structure and content of the data that determine LISP program flow. The implications of these characteristics of list processing languages for knowledge representation have shown themselves to be considerable.[37]

3.4.2 Expert Systems

The utility of AI technology for the modeling and simulation of complex systems has been demonstrated across a diverse range of applications. At the center of this activity is a burgeoning sub-field of AI, "expert systems," which is concerned with the construction of computer systems designed for heuristic problem solving. Indeed, Herbert Simon has suggested "heuristic search" as the central paradigm underlying AI in general.[38] Expert systems are specifically designed to assist people engaged in the business of problem solving.[39] The functional applications for expert systems include monitoring, control, interpretation, prediction, diagnosis, design, planning, debugging, repair, and instruction.[40] The substantive applications for this technology seem at present to be limited principally by imagination. The symbol manipulation approach of expert systems constructed in medicine, chemistry, military science, engineering, and psychology has permitted advances not possible with conventional quantitative modeling media.[41]

A few examples will illustrate the flexibility of expert system applications. DENDRAL is used in the identification of organic compounds.[42] PROSPECTOR is employed as a consultant in mineral prospecting.[43] SPILL helps workers at the Oak Ridge National Laboratory to locate and identify the sources and compositions of chemical spill accidents.[44] Rand's APM (advanced penetration model, also known as SWIRL, for simulating warfare in the ROSS language) is used by the U.S. Air Force to plan efficient penetration by strategic forces in a nuclear war with the Soviet Union. SWIRL is under further development to serve real-time command and control functions for the dynamic reprogramming of strategic forces *in bellum*.

An extended example of a successful expert system application is also instructive. A paradigmatic success story is provided by artificial intelligence research in the problem domain of medicine. Physicians face conflicting syndromes of symptoms. They must deal with large numbers of variables, which threaten explosive proliferations of possible solutions. In this respect the problem domain of medicine is typical of those domains in which expert systems have performed well.[45]

A pioneering expert system called MYCIN was developed partly as an experiment in artificial intelligence and partly as a diagnostic tool for doctors.[46] The substantive application of MYCIN is as an antimicrobial therapy consultant for physicians. In the interactive mode, MYCIN asks questions about the symptoms of the patient. The program then searches for matching information in its database until it identifies a possible or probable syndrome, finally recommending an antimicrobial therapy regime. For example, given a certain set of symptoms, MYCIN might predict an 80% probability that the problem is a staphylococcus infection and recommend a regime of thirty mils tetracycline three times daily for one week. Experiments pitting MYCIN against expert human physicians show it performing nearly as well, and in some cases, better than the human experts.[47]

Now it gets more interesting. The developers of MYCIN identified several problems with their creation. One of the most serious was the dependence of MYCIN upon its knowledge base, which in turn depended upon the people who wrote the computer program--mostly computer programmers, people who don't spend their spare time studying medical books. The programmers know all about machines but not so much about communicable diseases. To solve this and other problems, a "front-end" was designed for MYCIN, so that now MYCIN learns. The learning system is called TEIRESIAS.[48]

In theory the principal design feature of TEIRESIAS is a capability for the program to absorb the knowledge of a large community of human experts--in this case, physicians. The goal is to extract information from "conversations" between the physicians and the program and automatically add it to the knowledge base of the expert system in a form useful for making inferences about infectious disease diagnosis and therapy. To the extent that the machine is successfully able to interrogate human experts and capture within its circuits the essence of their domain expertise, the machine will accumulate the expertise of the human experts it has "talked to." Ultimately the machine will possess more technical knowledge and diagnostic heuristics related to problem domain of infectious disease therapy than any given human expert. The possibilities for applications of this technology in the area of military affairs is an awesome thing to contemplate, especially in an era when budget requests for research on "autonomous" unmanned combat robots are annually submitted to Congress.

3.4.3 Formalisms

Driving the growth in the number of successful expert system applications, there has been a steady improvement in the very high level list processing languages. These are themselves often implemented or written in LISP. Two important types of high-level list processing languages can be classified as the "rule-based" (or "procedural") and the "object-oriented" (or "declarative") languages, though recent developments have blurred this distinction considerably.[49]

ROSIE (for rule-oriented system for the implementation of expertise) is a good example of rule-based production system approaches to high-level list processing languages.[50] The basic building block of this type of language consists of "condition-action" rules of the form:

IF [condition] THEN [action1] ELSE [action2].

Concatenation of variations on this simple rule structure affords considerable symbolic power for knowledge representation. The SPILL program mentioned above was written in ROSIE.

A second common approach to high-level list processing languages can be referred to as the object-oriented message-passing languages. The basic building blocks of these languages are "objects." Objects are defined as individuals or as members of classes, each with its own database containing attributes and actions. Languages such as SMALLTALK,[51] PLASMA,[52] DIRECTOR,[53] and ROSS[54] support these capabilities. ROSS (for rule-oriented simulation system) is a good example. Developed at the Rand Corporation, ROSS adds to the flexibility of earlier object-oriented languages by allowing objects to have their behavior defined in terms of condition-action rules (conditionals or productions). The APM/SWIRL program mentioned above was written in ROSS.

The high-level list processing languages, both the rule-based languages like ROSIE and the object-oriented languages like ROSS, are part of a revolution in the computer sciences. The ultimate implications of the list processing languages for modeling and simulation have yet to be fully explored and, in most fields of inquiry--including political science and international relations--we have scarcely begun to exploit their known potential. The number of AI-inspired works in international relations research is small but growing.[55] An example application bearing on the problem at hand, that is, the role of strategy in arms accumulation, will suggest some of the possibilities.

3.4.4 Illustrating the Potential

Using a list processing language, such as an object-oriented message-passing simulation medium, the modeler has the capability of specifying variable attributes and behaviors. These attributes and behaviors can be seen as relations, because they are meaningful only in terms of their relationship to other objects in the system. A general form for defining such an object is:

```
10  Ask system to create a prototype <object-name>
20  with <attribute> <value> ...
30   and <memory> <value> ...
40   and <functions> <value> ...
50   and <plans> <value> ...
```

In this "rule," line 10 instructs the system to define an entity named "object-name." Line 20 asks the system to assign a list of attributes to the object, with values either given or variable. Actually, lines 30, 40, and 50 also can be seen as requests to the system to assign more attributes to the object, but these additional attributes have special functions so they get special names. "Memory" (line 30) comprises the knowledge base belonging to the object, containing all that the object knows about itself, its history, and its environment. "Functions" (line 40) comprises the behaviors of the object, defined in terms of the messages it sends to itself and to other entities in the environment. "Plans" (line 50) is a list of behaviors to be executed at some point in the future. Behavior options may include commands as flexible as: TELL <object> TO REMEMBER <nil>; or, WHEN RECEIVING <FOO> ASK YOUR <BAR> TO ADD <DILLI> TO ITS LIST OF <GAF>; or, PLAN AFTER <N> SECONDS TO INCREMENT YOUR <DOODAH> BY <VALUE>. These symbol handling capacities, then, represent arbitrary data structures, and as such, they are well suited to the modeling of dynamic interactive systems.

One illustration of the data formating capacities of this type of representation is to show how the substance of military strategy might be rendered into a machine manipulable form using such arbitrary data structures. To do so, we first define an object, NCA (National Command Authority), which has an attribute called a "plan." The plan, a subject-predicate template created with syntactic pattern-matching functions, is formally defined as:

$$\{p_1,p_2,p_3,p_4,p_5\},$$

where

p_1 is the definition of the situation
p_2 is the organization's objectives
p_3 is the designated set of means
p_4 is the specification of the instruments
p_5 is the specific plan for implementation

For the generic object NCA, the expanded form of the plan is:

{(The situation is that)(p_1),
(The object to)(p_2),
(The forces in being must)(p_3),
(The requirement is) (p_4),
(The plan is)(p_5)}.

For example, one instance of the plan for NCA might be as follows:

> {(The situation is that)(there is a basic conflict between the United States and the Kremlin),(The objective is to)(possess superior overall power for a policy of gradual coercion),(The forces in being must)(be able to conduct offensive operations to destroy vital elements of the Kremlin war-making capability),(The requirement is)(1,000 warheads deliverable to selected military-industrial targets),(The plan is) (distribution of platforms and targets)}.

This is not a hypothetical example. This code symbolically represents U.S. strategic military doctrine circa 1950.[56]

Such an approach has distinct advantages compared to conventional simulation approaches. Because it is composed more nearly of English-like statements, domain experts and those not versed in computer programming should find such models much more flexible--i.e., intelligible and modifiable--than one composed of quantitative mathematical statements. These advantages obtain at the output end, too, because selective model-based "meta-knowledge" (i.e., self-reporting and self-tracing functions) is often available.[57] This will impact on the reliability and risk estimation factors so often plaguing the acceptance of simulation modeling results. "In contrast to typical statistical approaches," as Thorson and Donald Sylvan note, the application of AI techniques "provides users (including, potentially, policy makers) with the opportunity to ask of the simulation decision aid, 'What can I do, and what is likely to happen if I do it?'"[58] This range of flexibility has obvious implications in terms of knowledge representation capabilities and possibilities for theory development and policy analysis.

An important advantage of the approach suggested here is that it encourages a rigorous focus on adaptive, purposive behavior. In traditional Richardson models, as well as simulations such as SIPER, SIMPEST, and GLOBUS, the output is in the form of time trajectories on at most a few descriptive variables, e.g., changes in levels of arms expenditures. In contrast, using the more powerful modeling-oriented languages such as those illustrated here, it is possible to model adaptive behavior. "In more advanced simulation studies, where . . . the model is an adaptive one, the model configuration may change throughout [the] simulation run. The output of the simulation may then include, in addition to the traditional numerical results, a documentation of the change in model composition to display its adaptive behavior."[59] Application of this modeling technology in the social sciences in general and peace science in particular warrants further investigation. The Richardson Tradition of armaments analysis seems like a good place to start, or more accurately, to play catch up, for other disciplines closely related to the Richardson Tradition have availed themselves to this technology with impressive results, as we shall see in the next section.

3.5 The Political Economy of Arms Modeling

This treatment has focused principally upon a class of models known as "arms-building models" and has generally neglected another closely related class of models often referred to as "arms-using models."[60] Most reviewers of the Richardson Tradition include as falling within that literature the arms-using type of model, also known as "Lanchester combat" or "dynamic attrition" models.[61] I have avoided this because the Lanchester models constitute, in some fundamental sense, a rival tradition to the Richardson Tradition. The general ethos underlying the combat models is a concern with winning. While this may or may not be a good thing to be concerned with, and while it *is* certainly a concern of those actually employed in making war, it was not Richardson's concern. As Richardson noted, his was not "a theory about victory and defeat," but an examination that attempted to take a scientific view of the phenomena of armaments accumulation and war, without regard to advantage.

Work on arms-using models has been very lively in recent years, more lively than the Richardson Tradition in many ways. Researchers of arms-using models can boast some very satisfying research results and, certainly, a higher level of research funding than work on arms-building models can claim. Part of the reason for the recent revival and success of arms-using models has to do with the employment of up-to-date methods. It may not be possible to emulate the success of research on arms-using models simply by using similar techniques on arms-building models, but those techniques do offer some interesting possibilities. This section begins to treat these possibilities.

3.5.1 SWIRLing to World War III: Technology and Models of War

A case study in the political economy of software development for arms analysis is provided by the Information Sciences Department of the Rand Corporation, a quasi-private research institute mainly involved in research for the U.S. Air Force. In particular, the story of two software development projects at Rand Information Sciences is relevant to this discussion. By the mid 1970s, the "expert system" phenomenon had begun to sweep over the computer science research institutes across the country. Artificial intelligence labs at MIT, at Carnegie-Mellon, at Xerox in Palo Alto, and elsewhere had begun to make big names for themselves by developing artificial intelligence-based systems of increasingly impressive capabilities. Most impressive of all, perhaps, were what we might call "meta-systems," developmental software engineering environments designed to assist and support the design of expert systems.[62] Meta-systems for AI were the cutting edge, the avant garde, the prestige project in that time--everybody who was anybody in AI was doing it. The leaders at Rand Information Sciences believed that scientific and other opportunities lay along this path. Thus was ROSIE conceived.

ROSIE is a general purpose, rule-based environment for the design of expert systems.[63] Implemented in the INTERLISP language, ROSIE's principal design goal was that the code should be English-like and thus highly readable by the "domain specialist," who may not have the machine-related expertise of the "knowledge engineer." This would facilitate the use of the programming tool by those domain specialists whose domain-specific knowledge must be encoded in order to construct the knowledge-based expert system. The "expert" part of expert systems, of course, comes not so much from the structure of the software as from the knowledge possessed by the human experts whose expertise is "mined" and inserted in the computer program.

In the development of large-scale computer programs, a prototype problem of the sort the system is being designed to work with is chosen as the first test case. In the case of ROSIE, the initial "bootstrap" domain problem-solving task was the so-called Oak Ridge National Laboratory (ORNL) "spill problem." The ORNL is a large, government-run research facility, located in three separate valleys in the Oak Ridge Mountains of Tennessee. The work at ORNL involves the use of many toxic substances, and there is considerable danger from accidents of various sorts; this is one of the reasons the lab is located in three separate valleys, lest no more than a third of the lab be lost in any particularly bad accident. The lab through all three valleys is laced with a series of storm drainage channels, all of which finally run together and drain into White Oak Creek. The ORNL spill problem is the problem of determining the identity and locating the source of pollutant substances found in the ORNL drainage system. It is a formidable problem given the extent of branching in the lab's drainage network and the wide variety of toxic and non-toxic substances that might be spilled.

A very large computer program consisting of about 500,000 lines of conventional code had been produced to assist workers at ORNL in dealing with the spill problem. It was slow, difficult to "maintain" (that is, keep up to date with the latest information), and "buggy" (i.e., it didn't always work well). The Rand knowledge engineers redesigned the spill problem program logic, rewriting the program using the ROSIE language. In all respects, the new version was superior, thus demonstrating the value of ROSIE.[64]

As ROSIE reached the end of its development phase, a problem became apparent. Though funded by federal agencies as a research project rather than as a commercial product, hopes for recouping any of the institutional overhead costs depended upon the strength of the market for large-scale general expert systems design environments. If nobody "bought" ROSIE, there would be minimal prestige benefits for Information Sciences. The market was not big.

Such observations help explain what emerged as a new strategy of software development at Rand. The next major AI software engineering project undertaken by the Information Sciences Department--ROSS--was developed from the start with a very specific market in mind: the Pentagon.[65] It was apparent to the knowledge engineers at Information Sciences that expert systems technology could be usefully applied to many of the sorts of problems of concern to the military. If this usefulness could be properly demonstrated to the right people, then adequate funding to sustain Information Sciences' foray into AI systems would likely be made available. Thus, the "bootstrap" domain problem to be worked on during the development of the ROSS software would be one known to inspire some interest in the Pentagon: the conduct of strategic nuclear warfare.

Following the pattern shown with ROSIE in using the existing SPILL program as the model for the domain bootstrap problem, ROSS's designers chose to emulate a rusty old Rand air-battle simulation called "ABM" (air battle model). Critics of the ABM simulation had pointed to numerous problems with the model, problems typical of this class of simulation models:

> The enormous amount of data required by ABM presented serious problems in gathering and validating data, as well as basic conceptual problems. As with other very large MSGs (Military Simulation Games), the problems with ABM highlight the need for sensitivity analysis, outside professional review, and even with simple data checking. It requires a leap of faith for anyone beside the model builders, who know what is in the computer programs that they have constructed, to apply and use the output from such simulations.[66]

The modelers at Rand called their new model APM (advanced penetration model). This connection with ABM was somewhat muted when they later changed the name of the model to SWIRL. The designers of APM (a.k.a. SWIRL) responded directly to the criticisms of ABM:

> Military battle simulations have been used extensively by the Air Force... They oftentimes lack the understandability and flexibility required by military planners and analysts. These simulations contain knowledge about the various objects being modeled including how the objects behave in various situations, how they interact with one another, and how they make decisions. Typically such knowledge is buried in unintelligible code that also lacks adequate documentation. Simulations would be even more valuable if users could better understand, and be able to modify, the behavioral knowledge embedded in such simulations. Analysts could better experiment with alternative strategies and tactics for both offensive and defensive forces.[67]

As the knowledge engineers at Rand put it, "To demonstrate the potential and payoff of the ROSS language, a prototype air penetration simulation, called SWIRL, was built."[68] SWIRL, which stands for simulating warfare in the ROSS language, is a game, but it is a very special kind of game. Actually, the algorithms employed to generate the mechanics of the air battle simulation are not based upon anything unique to AI technology, consisting as they do principally of chained probabilities and Bayesian logic. The object-oriented nature of the global simulation control structure, however, allows the complexity of interaction that makes the SWIRL simulation so powerful, and the user-oriented qualities of the front-end make that power available to users.

All objects in SWIRL are defined, in the first instance, in terms of a generic entity. From this original generic entity, a few attributes are added to make a particular class of generic objects. When a simulation run begins, the simulated objects are generated from appropriate template-like generic objects. To get a flavor of how this works, and what the code looks like, consider the following SWIRL routine.[69] This command instructs a particular fighter aircraft (itself derived from a generic fighter, which was in turn derived from a generic aircraft, and so on) to perform behaviors related to intercepting penetrating enemy aircraft (note that GCI stands for ground control intercept):

```
(ask FIGHTER when receiving
(chase >penetrator guided by >GCI)
(~you unplan all (land))
(~you set your status to scrambled)
(if (~you are on the ground) then
(~you take off))
(~requiring (~your guide-time) tell
~the GCI guide ~yourself to ~the penetrator)).
```

SWIRL is referred to as a "prototype air battle simulation." The particular sort of air battle played out in SWIRL is an offensive strategic attack by the U.S. Air Force against the Soviet Union. Among the objects defined in the SWIRL environment are various types of cargo, fighter, reconnaissance and bomber aircraft, vehicles such as the AWACs, enemy C^3I assets, command centers, airbases, radar sites, missile sites, and so on. All the weapons systems that would be involved in strategic nuclear war conducted by the U.S. Air Force against the Soviet Union, then, are represented as objects. The objects are defined in terms of their capabilities and characteristics (range of platforms and radars, kill probabilities against various targets, vulnerability probabilities against various defenses and offenses, etc.).

In order to execute a run of the SWIRL simulation, then, the user has only to specify the characteristics of the respective sides' hardware inventory and then to define a battle plan for the attack. SWIRL computes all the interactions, as well as the results of the attack in terms of what ends up getting destroyed and what is remaining. The idea is first of all to be able to simulate a wide variety of battle plans in order to find the best approach to such a strategic nuclear offensive. The secondary idea behind this, and the second phase of research on the ROSS-SWIRL project, was to redesign the system so that it might be useful for the dynamic reprogramming of forces remaining after an initial attack, so as to optimize regrouping for continuing the action against any targets that survive the initial phase of the war. This fits very nicely with the practical strategic plans expressed in Defense Secretary James Schlesinger's NSDM 232 and Defense Secretary Harold Brown's PD 59, and reaffirmed at the beginning of the Reagan administration, with Defense Secretary Casper Weinburger's expressed determination to beef up U.S. strategic C^3I resources in order to sustain a strategic nuclear war beyond the initial blows.

SWIRL had an important hardware element, as well, in the advanced methods employed to display the unfolding of the simulation game. Behind highly secured doors in the Rand Information Sciences ROSS lab, and occupying an entire wall of the large room, was an illuminated projection map of European Russia. Small symbols clustered here and there all over the map marked the location and identity of Soviet "assets": There were symbols for radar sites, with shading to show the radar coverage; symbols for Soviet tactical air force sites, containing symbols for Soviet air defense interceptor aircraft; and so on. A humorous Rand knowledge engineer even included little cross symbols to note the location of Soviet churches. There also is a complete set of symbols to indicate the location and activity of the attacking U.S. forces.

A simulation run on SWIRL with the wall display is an awesome thing to see. All of the symbols on the wall display are dynamic, and when a run is executed, the symbols on the display screen go into action. The interaction of weapons systems during strategic nuclear war in Soviet Russia is played out on the screen like a movie. The display is very "cute"; for example, when a nuclear weapon detonates in the course of a simulated battle, there is a little flash on the display at the target point, and the target symbol is then updated to identify its new status. Watching a full simulation run unfold on the wall display seems rather like what it would be to watch the actual fighting of World War III from the vantage point of a space station geo-synchronously "parked" over the northern plains of Europe. It is chilling.

The knowledge engineers at Rand Information Sciences were proud of the SWIRL wall display. They beamed when demonstrating to visitors a sample run of the war simulation on the newly up-and-running system. But they knew that the wall display was really nothing more than "bells and whistles," as far as the artificial intelligence researchers of the world were concerned. The real achievement, the technical triumph of ROSS and SWIRL, was in the control structures of the software, the utility of the object-oriented, rule-based, message-passing system. Beyond its function as an advanced output device, the bells and whistles were a public relations device, a sophisticated marketing tactic for ROSS and Rand's AI research. You could almost hear the thought, "When the boys over in war plans division see this . . ." Still, the intricate display is exciting more than anything else, in the final analysis, because it baldly displays the technical usefulness of this ROSS tool for the purpose at hand, planning strategic nuclear war. With such a model it will be possible to systematically explore the effectiveness of far more variations on offensive and defensive tactics than was previously conceivable. This is a gaming tool to blow the souls out of the adversary, to paraphrase Thomas Carlyle. The technical rational ego verily burbles at SWIRL. It is so wickedly beautiful. And it all started from a little arms-using type tactical battle simulation model. Wouldn't it be nice if we could do something like that for the arms-building models of the Richardson Tradition?

The contributions of Norbert Weiner to the field of cybernetics have led some to call him the father of modern inertial guidance systems, a key component making possible intercontinental strategic missiles. Though he was a jolly person, he waned to darkness when he discussed war gaming. Our programmed war games are the only available experience to fall back on considering nuclear war, because unlike, say, a counter-insurgency war, we cannot afford a few practice runs of nuclear war. Thus, our only way to experience a nuclear war safely is with programming games. "If we are to use this experience as a guide for our procedure in a real emergency, the values of winning which we have employed in the programming games must be the same values which we hold at heart in the actual outcome of the war. We can fail in this only at our immediate, utter, and irretrievable peril."[70]

3.6 Summary and Conclusions on Methods

The chapter can be summarized briefly:

§ *The structures, both conceptual and mathematical, used to represent data determine how it can be manipulated.*

§ *The methods of social physics alone may be insufficient to resolve all the problems of peace research.*

§ *The data structures of interest to the Richardson Tradition often do not conform to the quantitative metric.*

§ *The structures controlling changes in armaments accumulation processes remain ill defined.*

§ *Methods for representing planning processes deserve careful investigation as candidate structures.*

In the final analysis, the most challenging problems currently facing the Richardson Tradition of armaments analysis lie in the relationship between theory and method. The history of the sciences suggests that "theoretical and methodological concerns must go hand-in-hand. Progress in one area without similar advances in the other cannot be more than minimal."[71] There is more to science than technique. Blalock reminds us that "unless very careful attention is paid to one's theoretical assumptions and conceptual apparatus, no array of statistical techniques will suffice."

Indeed, more than twenty-five years ago, in the first major academic critique of Richardson's work, Anatol Rapoport argued the same case:

> The most intricate relations among various geometric figures, the most startling constructions with ruler and compass, the ultimate in ancient mathematical skill, would not be of any use in problems concerned with the spread of heat through a body or with what goes on in an alternating current circuit. On the other hand, statements such as "the rate of change of the sine function is the cosine function" . . . are entirely beyond the scope of ancient mathematics. They had no language in which to express such ideas. This is what we mean when we say that ancient mathematics is not applicable to classical physics. . . . Now let us turn the tables on classical analysis, and ask, "To what extent is classical analysis . . . applicable to theories of human behavior?" Is it possible that it lacks the necessary concepts, in the sense that ancient mathematics lacks certain concepts necessary for physical theory? . . . if the weakness of the method is in the paucity or in the irrelevance of its fundamental concepts, then resolution of technical difficulties will be of no help.[72]

The challenges facing researchers in the Richardson Tradition of armaments accumulation analysis consist not only in theory and/or in method, but in the relationship between the two. It seems entirely possible that the barriers to progress in the tradition, though apparently formidable, may yet yield under the persuasion of methods and theories not yet tried. The next chapter constitutes just such an attempt.

Notes

1. Herbert Simon, The Sciences of the Artificial (Cambridge, MA: MIT Press, 1969, 1981).

2. G.A. Miller, E. Galanter, and K.H. Pribram, Plans and the Structure of Behavior (New York: Holt, Rinehart and Winston, 1960), p. 213.

3. Peter Winch, The Idea of a Social Science (London: Routledge and Kegan Paul, 1958).

4. Ibid., p. 74.

5. W. Ross Ashby, Design for a Brain: The Origin of Adaptive Behavior (New York: John Wiley and Sons, 1960), p. 62.

6. Lewis Fry Richardson, Arms and Insecurity: A Mathematical Study of the Causes and Origins of War (Pittsburgh: Boxwood Press, 1960), p. xiii.

7. Gregory Bateson, "Culture Contact and Schismogenisis," Man 35 (1935), pp. 173-183.

8. Miller, Galanter, and Pribram, Plans and the Structure of Behavior, op. cit.

9. Hayward Alker, Jr., "From Political Cybernetics to Global Modeling," paper presented to the annual meeting of the American Political Science Association, New York, August 31 to September 3, 1978, p. 7.

10. Gregory Bateson, "The Pattern of an Armaments Race--Part I: An Anthropological Approach," Bulletin of the Atomic Scientists 2 (1946), pp. 10-11; and also his "The Pattern of an Armaments Race--Part II: An Analysis of Nationalism," Bulletin of the Atomic Scientists 2 (1946), pp. 26-28; cf. Lewis Fry Richardson, Generalized Foreign Politics (Cambridge, England: Cambridge University Press, 1939).

11. Max Weber, "The Interpretive Understanding of Social Action," in Readings in the Philosophy of the Social Sciences, ed. May Brodbeck (New York: Macmillan, 1968), p. 24.

12. Irme Lakatos, "Falsification and the Methodology of Scientific Research Programmes," in Criticism and the Growth of Knowledge, eds. Irme Lakatos and Alan Musgrave (London: Cambridge University Press, 1970), pp. 91-196.

13. Georg Henrik Von Wright, Explanation and Understanding (Ithaca, NY: Cornell University Press, 1971).

14. In the sense of Thomas S. Kuhn, The Structure of Scientific Revolutions, 2d ed. (Chicago: University of Chicago Press, 1970).

15. Ibid.

16. Hayward R. Alker, Jr., "Research Paradigms and Mathematical Politics," paper presented to the International Political Science Association Meeting, Mannheim, Germany, 1971.

17. Richard K. Ashley, "Noticing Pre-paradigmatic Progress," in In Search of Global Patterns, ed. James N. Rosenau (New York: Free Press, 1976), pp. 150-158.

18. Von Wright, Explanation and Understanding, op. cit.

19. N. Rashevsky, Mathematical Theory and Human Relations, An Approach to a Mathematical Biology of Social Phenomena (Bloomington, IN: Principa Press, 1947); Anatol A. Rapoport, "Lewis Fry Richardson's Mathematical Theory of War," Journal of Conflict Resolution 1 (1957), pp. 249-299; Thomas L. Saaty, Mathematical Models of Arms Control and Disarmament: Applications of Mathematical Structures in Politics (New York: Wiley and Sons, 1968); Claudio A. Cioffi-Revilla, "Theories of Arms Races and Mathematical Structures in International Relations," paper presented to the joint meeting of the Mid-west Political Science Association and the Mid-west section of the Peace Science Society (International), Chicago, April 19-20, 1979; Philip A. Schrodt, "Artificial Intelligence and the State of Mathematical Modeling in International Relations," paper presented to the Conference on Dynamic Models of Conflict, Boulder, CO, October 31 to November 3, 1984; Charles F. Anderton, "Arms Race Modeling: Categorization and Systematic Analysis," paper presented to the annual meeting of the International Studies Association, Washington, DC, March 5-9, 1985.

20. Cioffi-Revilla, "Theories of Arms Races and Mathematical Structures in International Relations," op. cit., p. 67.

21. Georg Wilhelm F. Von Leibnitz, The Monadologie, trans. H.W. Carr. (Los Angeles: University of Southern California, 1930).

22. Simon, The Sciences of the Artificial, op. cit.

23. Robert G. Wesson, "Non-spending in Secret for Defense: Big Budgets Impress the Soviets, So Dummy Outlays Could Help," Los Angeles Times, February 15, 1985, Part II, p. 5.

24. Dagobert Brito and Michael D. Intriligator, "Some Applications of the Maximum Principle to the Problem of an Armaments Race," in Modeling and Simulation, eds. William G. Vogt and Marlin H. Mickle (Pittsburgh: Instrument Society of America, 1973), p. 140.

25. Richard J. Stoll, "Let the Researcher Beware: The Use of Richardson Equations to Estimate the Parameters of a Dyadic Arms Acquisition Process," American Journal of Political Science 26 (1982), p. 88.

26. See Table 4.1 on p. 178 in Niklaus Wirth, Algorithms and Data Structures (Englewood Cliffs, NJ: Prentice-Hall, 1986).

27. See, for example, Dwain Mefford's work, including "Using the Past to Create the Future: The Role of Cognition and Memory in Crisis and War," paper presented to the annual meeting of the International Studies Association, Atlanta, March 27-31, 1984; and also Mefford's "Formulating Foreign Policy on the Basis of Historical Analogies: An Application of Developments in Artificial Intelligence," paper presented to the annual meeting of the International Studies Association, Atlanta, March 27-31, 1984; and his "Behind Every Solution Is a Story: Story Grammars and Extended Game

Theory," paper presented to the annual meeting of the International Studies Association, Washington, DC, March 5-9, 1985; and also his "Changes in Foreign Policy Across Time: The Logical Analysis of a Succession of Decision Problems Using Logic Programming," in Dynamic Models of International Conflict, ed. Urs Luterbacher and Michael D. Ward (Boulder, CO: Lynne Reinner Publishers, 1985), pp. 401-423; also see Noel Kaplowitz, "Psychopolitical Dimensions of International Relations: The Reciprocal Effects of Conflict Strategies," International Studies Quarterly 28 (December 1984), pp. 373-406; and also James P. Bennett, "Data Stories: Learning About Learning from the U.S. Experience in Vietnam," in Foreign Policy Decision Making: Perception, Cognition, and Artificial Intelligence, ed. D.A. Sylvan and S. Chan (New York: Praeger, 1984), pp. 227-279.

28. Schrodt, "Artificial Intelligence and the State of Mathematical Modeling in International Relations," op. cit., p. 7.

29. A.D. Aleksandrov et al., eds., Mathematics: Its Content, Methods, and Meaning, 2d ed. (Cambridge, MA: MIT Press, 1984), 3 vols.

30. R. Miller and Stuart Thorson, "Control Concepts in Theories of Governmental Decision-Making: A Structural Approach," in Mathematical Systems in International Relations Research, ed. John V. Gillespie and Dina A. Zinnes (New York: Praeger, 1977), pp. 60-90.

31. Ibid., p. 89.

32. Paul A. Anderson and Stuart J. Thorson, "Artificial Intelligence Based Simulation of Foreign Policy Decision-Making," Behavioral Science 27 (April 1982), pp. 176-193; Stuart J. Thorson and Donald A. Sylvan, "Counterfactuals and the Cuban Missile Crisis," International Studies Quarterly 26 (December 1982), pp. 539-571; Stuart J. Thorson, "Intentional Inferencing in Foreign Policy: An AI Approach," in Foreign Policy Decision Making: Perception, Cognition, and Artificial Intelligence, ed. Donald A. Sylvan and Steve Chan (New York: Praeger, 1984), pp. 280-309; and also Thorson's "The Intentional Inferencing Problem in Foreign Policy Decision Making," paper presented to the annual meeting of the International Studies Association, Washington, DC, March 5-9, 1985; Stuart J. Thorson and Kristi Andersen, "A Conceptual Model of Intentional Inferencing in Foreign Policy," paper presented to the annual meeting of the International Studies Association, Anaheim, CA, March 25-29, 1986.

33. Terry Winograd, "Frame Representation and the Declarative-Procedural Controversy," in Representation and Understanding, ed. Daniel Bobrow and A. Collins (New York: Academic Press, 1975), pp. 185-210.

34. Charles W. Ostrom, Jr., "Evaluating Alternative Foreign Policy Decision-Making Models: An Empirical Test between an Arms Race Model and an Organizational Politics Model," Journal of Conflict Resolution 21 (June 1977), pp. 244, 245.

35. Herbert Simon, "Artificial Intelligence Research Strategies in the Light of AI Models of Scientific Discovery," Proceedings of the sixth International Joint Conference on Artificial Intelligence, Tokyo, August 20-23, 1979, Vol. 2., pp. 1086-1094.

36. John McCarthy, P.W. Abrahams, D.J. Edwards, T.P. Hart, and M.I. Levin, LISP 1.5 Programmer's Manual (Cambridge, MA: MIT Press, 1962); E.C. Berkeley and Daniel G. Bobrow, The Programming Language LISP (Cambridge, MA: Information International, 1964); and their The Programming Language LISP: Its Operation and Applications (Cambridge, MA: MIT Press, 1966); C. Weissman, LISP 1.5 Primer (Belmont, CA: Dickenson, 1967); W. Teitelman, The INTERLISP Reference Manual (Palo Alto, CA: Xerox Palo Alto Research Center, 1974); P.C. Jackson, An Introduction to Artificial Intelligence (New York: Mason Lipscomb, 1974); M. King and P. Hayes, "An Introduction to Programming in LISP," in Computational Semantics, ed. E. Charniak and Y. Wilks (Amsterdam: North-Holland Publishing, 1976), pp. 235-282; Patrick H. Winston, Artificial Intelligence (Reading, MA: Addison-Wesley Publishing Co., 1977); Patrick H. Winston and B.K.P. Horn, LISP (Reading, MA: Addison-Wesley Publishing Co., 1981); Guy L. Steele, Jr., Common LISP (Bedford, MA: Digital Press, 1984).

37. Daniel G. Bobrow and A. Collins, eds., Representation and Understanding (New York: Academic Press, 1975); Daniel G. Bobrow and Terry Winograd, "An Overview of KRL, a Knowledge Representation Language," Cognitive Science 1 (1977), pp. 3-46.

38. Simon, "Artificial Intelligence Research Strategies in the Light of AI Models of Scientific Discovery," op. cit.

39. Frederick Hayes-Roth, Donald A. Waterman, and Douglas Lenat, eds., Building Expert Systems (Reading, MA: Addison-Wesley, 1983); and E. Edelson, "Expert Systems--Computers That Think Like People," Popular Science 221 (September 1982), pp. 48-52.

40. Hayes-Roth et al., eds., Building Expert Systems, op. cit.

41. Tuncer I. Oren and Bernard P. Ziegler, "Concepts for Advanced Simulation Methodologies," Simulation 32 (March 1979), pp. 69-82; Donald Michie, ed., Expert Systems in the Micro-Electronic Age (Edinburgh: Edinburgh University Press, 1979).

42. Edward A. Feigenbaum, "Themes and Case Studies of Knowledge Engineering," in Expert Systems in the Micro-Electronic Age, ed. Donald Michie (Edinburgh: Edinburgh University Press, 1979), pp. 3-25.

43. R. Duda, J. Gaschnig, and P. Hart, "Model Design in the PROSPECTOR Consultant System for Mineral Exploration," in Expert Systems in the Micro-Electronic Age, ed. Donald Michie (Edinburgh: Edinburgh University Press, 1979), pp. 153-167.

44. Frederick Hayes-Roth, "A Tutorial on Expert Systems: Putting Knowledge to Work," Proceedings of the seventh International Joint Conference on Artificial Intelligence, University of British Columbia, Vancouver, August 24, 1981.

45. E.H. Shortliffe, B.G. Buchanan, and E.A. Feigenbaum, "Knowledge Engineering for Medical Decision-Making: A Review of Computer-Based Clinical Decision Aids," Proceedings of the IEEE 67 (1979), pp. 1207-1224.

46. E.H. Shortliffe, Computer-Based Medical Consultation: MYCIN (New York: American Elsevier, 1976).

47. V.L. Yu, L.M. Fagan, S.M. Wraith, W.J. Clancey, A.C. Scott, J.F. Hannigan, R.L. Blum, B.G. Buchanan, and S.N. Cohen, "Antimicrobial Selection by Computer: A Blinded Evaluation by Infectious Disease Experts," Journal of the American Medical Association 242 (1979), pp. 1279-1282.

48. Randall Davis, "Applications of Meta-Level Knowledge to the Construction, Maintenance and Use of Large Knowledge Bases" (Ph.D. dissertation, Stanford University, 1976).

49. J. Mylopoulos, "An Overview of Knowledge Representation," SIGART Newsletter 74 (January 1981), pp. 5-12; N.V. Findler, ed., Associative Networks: Representation and Use of Knowledge by Computers (New York: Academic Press, 1979); Donald Waterman and Frederick Hayes-Roth, Pattern-Directed Inference Systems (New York: Academic Press, 1979); Bobrow and Collins, eds., Representation and Understanding, op. cit.; Terry Winograd, "Frame Representation and the Declarative-Procedural Controversy," in Representation and Understanding, ed. Daniel Bobrow and A. Collins (New York: Academic Press, 1975), pp. 185-210.

50. D. Gorlin, Frederick Hayes-Roth, Stan Rosenschein, and Henry Sowizral, The ROSIE Language Reference Manual N-1657-APRA (Santa Monica, CA: Rand Corp., 1982); also their Programming in ROSIE: An Introduction by Means of Examples N-1646-ARPA (Santa Monica, CA: Rand Corp., 1982); and also Rationale and Motivation for ROSIE N-1648-ARPA (Santa Monica, CA: Rand Corp., 1982).

51. A. Goldberg and A. Kay, SMALLTALK-72 Instruction Manual SSL 76-6 (Palo Alto, CA: Xerox Palo Alto Research Center, 1976).

52. C. Hewitt, "Viewing Control Structures as Patterns of Passing Messages," Artificial Intelligence 8 (1977), pp. 323-364.

53. K.M. Kahn, DIRECTOR Guide AI Memo 482-B (Cambridge, CA: MIT, Artificial Intelligence Lab, 1979).

54. Philip Klahr and William S. Faught, "Knowledge-Based Simulation," Proceedings of first meeting of the American Association for Artificial Intelligence, Stanford, CA, August 18-21, 1980, pp. 181-183.

55. See, for example, Hayward R. Alker, Jr., and C. Christensen, "From Causal Modeling to Artificial Intelligence: The Evolution of a UN Peace-Making Simulation," in Experimentation and Simulation in Political Science, ed. J.A. LaPonce and Paul Smoker (Toronto: Toronto University Press, 1972); Anderson and Thorson, "Artificial Intelligence Based Simulation of Foreign Policy Decision-Making," op. cit.; Howard Tamashiro, "Algorithms, Heuristics, and Artificial Intelligence Modelling of Strategic Statecraft," in Foreign Policy Decision Making: Perception, Cognition, and Artificial Intelligence, ed. Donald A. Sylvan and Steve Chan (New York: Praeger, 1984), pp. 197-226; Akihiko Tanaka, "China, China Watching, and CHINA_WATCHER," also in the Sylvan and Chan volume, pp. 310-344; other examples are several papers in the bibliography by James Bennett, Craig Etcheson, Dwain Mefford, Phil Schrodt, Stuart Thorson, and their collaborators.

56. Etcheson and Scacchi, "MONADYSIM: Applying AI Techniques to the Study of War," op. cit.

57. Randall Davis and Bruce Buchanan, "Meta-Level Knowledge: Overview and Applications," Proceedings of the fifth International Joint Conference on Artificial Intelligence, Cambridge, MA, August 22-25, 1977; and Avron Barr, "Meta-Knowledge and Cognition," Proceedings of the sixth International Joint Conference on Artificial Intelligence, Tokyo, August 20-23, 1979, Vol. 1, pp. 31-33.

58. Thorson and Sylvan, "Counterfactuals and the Cuban Missile Crisis," op. cit.

59. Tuncer I. Oren, "Concepts and Criteria to Assess the Acceptability of Simulation Studies: A Frame of Reference," Communications of the ACM 24 (1981), pp. 180-189.

60. Kendall D. Moll and G.M. Luebbert, "Arms Race and Military Expenditure Models," Journal of Conflict Resolution 24 (March 1980), pp. 153-185.

61. For example, see Martin C. McGuire, "Theoretical Notes on Lanchester Models of Military Tactics, Troop Density, Organization, and Attrition," paper presented to the Conference on Dynamic Models of Conflict, Boulder, CO, October 31 to November 3, 1984; and Charles F. Anderton, "Arms Race Modeling: Categorization and Systematic Analysis," paper presented to the annual meeting of the International Studies Association, Washington, DC, March 5-9, 1985.

62. Teitelman, The INTERLISP Reference Manual, op. cit.; Goldberg and Kay, SMALLTALK-72 Instruction Manual, op. cit.; Kahn, DIRECTOR Guide, op. cit.

63. Gorlin et al., The ROSIE Language Reference Manual, also Rationale and Motivation for ROSIE, and Programming in ROSIE: An Introduction by Means of Examples, all op. cit.

64. Hayes-Roth, "A Tutorial on Expert Systems: Putting Knowledge to Work," op. cit.

65. David McArthur and Philip Klahr, "The ROSS Language Manual," Report N-1854-AF (Santa Monica, CA: Rand Corp., 1982); also Philip Klahr, David McArthur, Sanjai Narain, and Eric Best, "SWIRL: Simulating Warfare in the ROSS Language," Report N-1885-AF (Santa Monica, CA: Rand Corp., 1982).

66. Garry D. Brewer and Martin Shubik, The War Game: A Critique of Military Problem Solving (Cambridge, MA: Harvard University Press, 1979), p. 131.

67. Klahr, McArthur, Narain, and Best, "SWIRL: Simulating Warfare in the ROSS Language," op. cit., p. v.

68. Ibid., p. iii.

69. Ibid., p. 7.

70. Norbert Weiner, Cybernetics, or Control and Communication in the Animal and the Machine, 2d ed. (Cambridge, MA: MIT Press, 1961), p. 177.

71. H.M. Blalock, Jr., Conceptualization and Measurement in the Social Sciences (Beverly Hills, CA: Sage, 1982), p. 9.

72. Anatol A. Rapoport, "Lewis Fry Richardson's Mathematical Theory of War," op. cit.

4 A COMPUTATIONAL MODEL OF ARMS ACCUMULATION

In Chapter Four, a "finite state machine" model of armaments accumulation is presented. It is called ARES, and is implemented in a micro-computer dialect of the LISP language. ARES differs from other simulations of armaments accumulation in its focus on the role played by strategies of conflict in bringing about qualitative changes in the force postures of nations in conflict. The spatial domain of ARES is modeled after the Hellenic world, and the temporal domain after the fifth century B.C.[1]

This chapter begins with a review of the simulate historical context of the ARES simulation system. Problems of modeling conflict situations and the need for alternative conceptualizations of arms accumulation, which motivate this effort, are then introduced. The focus, however, is on the model and the "generator" type of control structure it employs. The body of this chapter is devoted to an introduction to the ARES arms race simulation system, including a description of the model, the rationale behind it, a sample session, and experimental run. The chapter concludes with a discussion of the advantages and the limitations of models like the ARES armaments accumulation simulation.

4.1 Pseudo-Historical Background

Heroditus tells of a certain Athenian citizen, Themistocles, whose strategic vision was vital in preserving freedom for all Hellas.[2] After the defeat of the Persian emperor Darius at Marathon in 490 B.C., prevailing public opinion in Athens held that Persian tyranny was no more, and, as it were, "peace and freedom are at hand." Themistocles had his doubts, but he did not give them public voice. Instead, in a debate over the interpretation of an oracle from Delphi, he successfully argued that the gods bid the city to invest a fleet of one hundred ships to support Athens in its maritime rivalry with their cousins, the Eginetans.

The Roman historian Plutarch suggests that Themistocles had a hidden agenda. His true motivation in levying the fleet was that he "foresaw from far before what would be necessary" in the event the Persian hordes returned.[3] Athens had been lulled into a false sense of security by one particularly glorious battle, but Themistocles feared the power of the Persian tyrants, who could conscript huge armies. So he prepared for a strategy of sea control. Athenian naval supremacy, founded upon Themistocles' one hundred-ship fleet, would later prove pivotal in repelling the Persians under Xerxes, son of Darius, in the action at Salamis in 480 B.C. The fleet never saw action against the Eginetans.

The Athenian victory at Salamis marked the end of the Median War and left Athens with a large naval force. Being prudent people, the Athenians did not allow this asset to waste. The glory and reknown of Athens grew in proportion to the exploits of its seafarers throughout the Aegean Sea, and the exploits were legion. "The growth in the power of Athens," the historian Thucydides asserts, "and the alarm which this inspired in Lacedaemon, made war inevitable."[4] Thus, the Athenian fleet remaining after the Median War became one of the central strategic weapons systems in an even more pivotal conflict--the Peloponnesian War. This war, principally a conflict between the city-states of Sparta and Athens, made the preceding Median War seem tame. "Never had so many cities been taken and laid desolate . . . never was there so much banishing and blood-shedding. . . ."

The ancient observers report that an event of considerable consequence transpired, but they disagree on the causes of the event. Contemporary social scientists have joined these disputes, attemping to devise new methods for settling the arguments. One popular method involves building and testing mathematical models of the events.

4.2 Modeling Ancient Conflict

There are several ways to approach a theory that attempts to account for the apparent military interaction leading to the Peloponnesian War. One might hypothesize that a reaction process was at work here. The ruling oligarchy in Sparta may have been so alarmed at the military capability and aggressive policies of the Athenians that they increased public and private preparations for military action in advance of any possible trouble. In turn, Sparta's war-like preparation may have been perceived in Athens as an indication of hostility, accelerating still further the warlike preparations of the Athenians. If x and y represent military preparations in Athens and Sparta, respectively, dx/dt and dy/dt are changes in those preparations and A and B are constants representing their respective reactivity to changes in the other, then the hypothesized mutual reaction processes preceding the Peloponnesian War are described by the following equations:

$$dx/dt = Ay \qquad [4.1]$$

$$dy/dt = Bx \qquad [4.2]$$

On the other hand, an attempt to theorize about the armaments accumulation processes associated with the Peloponnesian War may take an entirely different tack. One could hypothesize that this accumulation of arms was principally driven by factors internal to the city-states. It is well known that Themistocles was an ambitious and exceedingly virtuous citizen, wont to regard glory as a supreme virtue. It may be that the ambitions of Themistocles and his ilk are principally responsible for the accumulation of military capability. This might even be reinforced by subtle economic pressures from the Athenian ship builders employed to construct the fleet or by the caste of adventurers and mercenaries thrown out of work at the end of the Median War. If one allows x, y, dx/dt, and dy/dt to represent the same things as in Equations

4.1 and 4.2, but now set C and D to represent the tendency of the respective city-states to invest in military capability, the hypothesized autonomous intertia processes in the Peloponnese are given by:

$$dx/dt = Cx \qquad [4.3]$$

$$dy/dt = Dy \qquad [4.4]$$

Employing either the mutual reaction hypothesis modeled in Equations 4.1 and 4.2 or the autonomous inertia hypothesis modeled in Equations 4.3 and 4.4 or both, we would be pursuing what has become the orthodox experimental design in the Richardson Tradition of armaments analysis.[5] Based on these models, we could perform stability and equilibrium analyses, gathering information to test Thucydides' assertion on the inevitability of the war. After data collection, selection of an estimator, and the calculation of a few test statistics (coefficient of determination, regression coefficients, and some measures of reliability), we would be prepared to test some experimental hypotheses and determine to what extent the predicted patterns appear to explain variances in the data and, with some degree of confidence, whether this is the product of random chance. Quite a lot of effort, altogether, to be sure. And still we would not know how or why they fought.

4.2.1 Things Missing in the Richardson Model

After all, we don't want to forget Themistocles. Clearly, Themistocles was what we would refer to as a "strategic thinker." He attacked the Persians' strategy. Unable to match Persia's manpower, the Athenians turned to superior technology. The fleet enabled the Athenians to take the fight to the invaders, defeat the empire, and then to cultivate their own far-flung alliance. The strategy of sea control was a response to a long-range potential threat rather than a present danger, and it had long-range consequences. It was a strategy of preparedness. This strategy guided public spending in Athens for much of the fifth century B.C., and the public spending priorities, in turn, shaped the Athenian military force posture. The relationships among strategic goals, the evolution of force posture, and the reciprocal impact of strategy and capability between city-states are not captured in a theory reduced to one of the models formulated above. And yet these things remain central to an understanding of armaments accumulation in the Peloponnese during the fifth century B.C., no less than they are central to the understanding of any particular instance of arms accumulation.

One of the long-range consequences of Athenian investment toward a strategy of sea control, long after the Persian threat had been dispelled, was to produce a reaction in neighboring city-states. The crucial institutional forces within each city-state that shape responses in such a situation, as Wohlstetter points out, are not reflected in action-reaction models of the Richardson Tradition.[6] Samuel Huntington and Colin Gray also discuss the importance of strategic planning in the armaments accumulation process.[7] Gray posits one such model, which he calls "absolute planning": "Suppose that a state observes the probable course of military policy of its rival, identifies an appropriate force structure and then builds according to its own blueprint--insensitive even to fairly large changes in the rival's programs. . . ."[8] And, perhaps, even insensitive to changes in the identity of the rival. The theoretical gaps that arouse these critics, and to which the present research attempts to respond, concern the relationship of strategy to continuity and change in the organization of military bureaucracies.

4.2.2 State Spaces and State Transformation

Representing a dynamic system such as an arms accumulation process reduces to a question of state transformation. That is, identifying the critical variables of interest in the system, one can describe the state of the system at any time by giving the values of those critical varibles. The set of all possible states of the system is called the "state space." The state of the system at the beginning of the time period under study is called the "initial state." Similarly, one can describe the end of the time period under study in terms of the final or "goal state." Thus, to describe the dynamic behavior of the system, one must identify a "control structure" that traces a path through the state space from the initial state to the goal state.

Figure 2.2 shows the state space typically explored in arms race studies, and Equations 4.1 to 4.4 show typical Richardson control structures. The state of the system at time t is given by the simple 2-tuple (x, y), where x and y have the same values as in the equations. The difficulty, then, lies in that this nomenclature simply was not designed to deal with such data as strategy and force posture. A single aggregate real number does not contain much information about force posture. The research reported here represents an attempt to fashion an approach that overcomes this limitation. List processing languages allow data structures, state spaces, and control structures that permit us to theorize more directly about such qualitative properties of armaments accumulation processes as strategy and force posture.

4.3 ARES

4.3.1 What Is ARES?

Ares is the Greek god of war. ARES stands for <u>a</u>rms <u>r</u>ac<u>e</u> <u>s</u>imulation. ARES simulates the strategic military interaction of Sparta and Athens prior to the outbreak of the Peloponnesian War. It was a war of such intensity, duration, and destructiveness, so the ancients tell us, that even the gods were involved in the fighting.

Within the ARES simulation environment, we encounter a world of two entities, MELOS and DELOS. The island of Delos was the location of the treasury of the Confederacy of Delos, a sort of United Nations of the Hellenic world, which became a front organization for Athenian imperial ambitions. Melos was a small, would-be "neutral" island nation populated by a just, proud, idealistic, and somewhat hapless people, on-again, off-again allies of the Spartans. Melos has been selected to represent the Spartan alliance principally for aesthetic reasons.

ARES is a test-bed for counter-factual hypothesis testing in the domain of strategic military interaction. It is not meant to be a true representation of the Hellenic world, though the simulate empirical results display a curious isomorphism with the historical record, suggesting a certain degree of face validity. This is an especially surprising result given the very simple and anachronistic strategic decision logic employed in ARES. ARES should be viewed as an heuristic exercise aiming at the introduction of a formalism that may apply to other cases.

4.3.2 Goal

ARES is the product of a larger research program, the underlying purpose of which is to explore new methodologies for the extension/synthesis of normative and empirical peace research. I have selected a theoretical focus on the dynamic adaptive behavior of large-scale organizations because I believe that here lie fruitful avenues of exploration in response to challenges within two broad traditions of international relations peace research: the Richardson Tradition of arms analysis and the emergent global modeling tradition. The former is defined broadly so as to include the game theoretic, control theory, operations analytic, and other innovations as well as the large-scale econometric modeling work by scholars such as Nazli Choucri and Robert North and Ashley. Likewise, we define global modeling in a non-restrictive sense, including the Guetzkow School (INS, IPS, SIPER, GLOBUS), the Geneva School (SIMSWISS, SIMPEST), the Club of Rome School (WORLD2, WORLD3, WIM), and other major projects such as MOIRA, SARU, FUGI, and LAWM, as well as those efforts of a more restricted scope (TEMPER, PRINCE, DPM, SWIRL, AEM, etc.). To a large extent, the Richardson Tradition and global modeling overlap; most of the models described above represent military interaction as a simple, additive Richardson-type reaction process. The substantive domain of the present effort lies within this region of overlap between the two traditions.

4.3.3 Domain

Sometimes in the past scholars have surrounded the puzzle of armaments accumulation with a bit too much mystery. One suspects that most armaments accumulation behavior can be traced to a frightfully simple cause. Arms races happen because that's the policy. It is all around us: We can't see the forest for the trees. This is so manifest that one courts accusations of tautology, but if we look to the sources of governments' policies in the military domain we are immediately delivered into a world of ideas about strategy, structured by bureaucracies that at once enable and constrain the realization of the ideas.

Theories about the role of strategy in armaments accumulation processes remain underdeveloped due to the fact that the dominant grammar of thought in the study of arms accumulation is not particularly conducive to the representation and manipulation of such qualitative social concepts as military strategy and bureaucratic behavior.[9] The relationship between strategy and structure in international military interaction constitutes the substantive domain of the simulation. ARES is modeled after a particular category of this substantive domain, that type found in the wars of Greece in the fifth century B.C., in such a way as to challenge the reductionist tendencies imposed by the quantitative methods.

4.3.4 Architecture

The ARES simulation system is an interactive environment for experimental exploration of strategic military behavior in the Peloponnesian War. It is user friendly and menu driven and can be modified or extended by the user with relative ease. Conceived principally as an heuristic exercise, ARES is useful beyond the fact that it forces the user to think about the relationship between strategy and structure in military interaction. It also suggests possibilities for further modeling efforts, such as contemporary Soviet-American interaction. For the present, however, we will restrict ourselves to a discussion of the structure of ARES.

ARES Conceptual Structure

Figure 4.1 illustrates the functions performed by the user, DELOS, and MELOS in ARES. The role of the user is to "play god" by defining the history of the world and the initial conditions in the two city-states. Thus, for each period in the simulation run the strategy of DELOS and the aggregate annual military expenditure of DELOS and MELOS are given as independent variables. DELOS "acts" by distributing its annual military budget among the three branches of its military structure (AIR, ARMY, NAVY) according to the plans embedded in its strategy. MELOS is more interesting because it is reactive. MELOS "observes" the forces fielded by DELOS and, correctly or incorrectly, makes an inference about the strategy behind those forces. MELOS then "rationally" selects an optimal counter-strategy from the universe of possible strategies and proceeds to distribute its own budget according to that plan.

ARES Object Structure and Behavior

ARES is a simple application in a style of simulation modeling called "object-oriented" programming. Here, we define a set of objects in terms of their attributes, behaviors, and other characteristics. The relationships among the objects generate the dynamics of the system. In ARES the objects are militarized city-states with five attributes: a strategy, army, navy, air force, and an annual budget (see Figure 4.2). The variables describe the state of the system at any time.

The behaviors of the city-states vary. The behavior of DELOS is limited to the capacity to change the structure of its military force posture over time in accordance with a given plan or strategy. The behavior of MELOS is more complex. In addition to the capacity to change its force posture as does DELOS, MELOS also has the capacity to adapt its strategy as conditions in the environment change. These behaviors are accomplished by means of the functions listed in Table 4.1. Two of the functions control the dynamics of strategic interaction between the city-states. The function called REACT controls change in MELOS, and will be discussed in section 4.4 and Appendix B. The function called SUPPLY is used by both city-states to distribute their resources among their military services. SUPPLY accomplishes this by matching the strategy of the city-state with a set of allocative priorities and then disbursing the aggregate annual budget according to the set of priorities. SUPPLY also will be discussed here as well as in the appendices.

4.3.5 ARES Operation

Input to ARES must be of the following form:

$$((\text{STRATEGY}_i\ X_i\ Y_i)(\text{STRATEGY}_{i+1}\ X_{i+1}\ Y_{i+1}) \ldots n)$$

where:

Figure 4.1 ARES Conceptual Structure

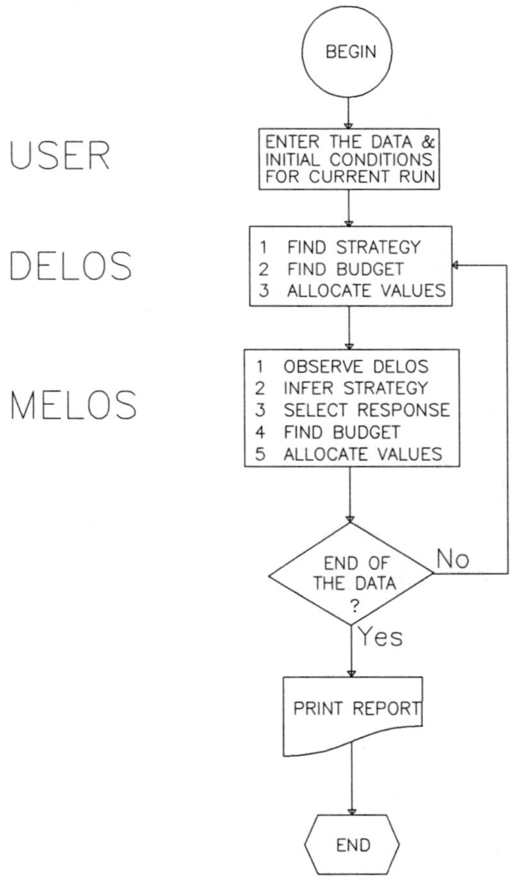

Figure 4.2 ARES Object Structure

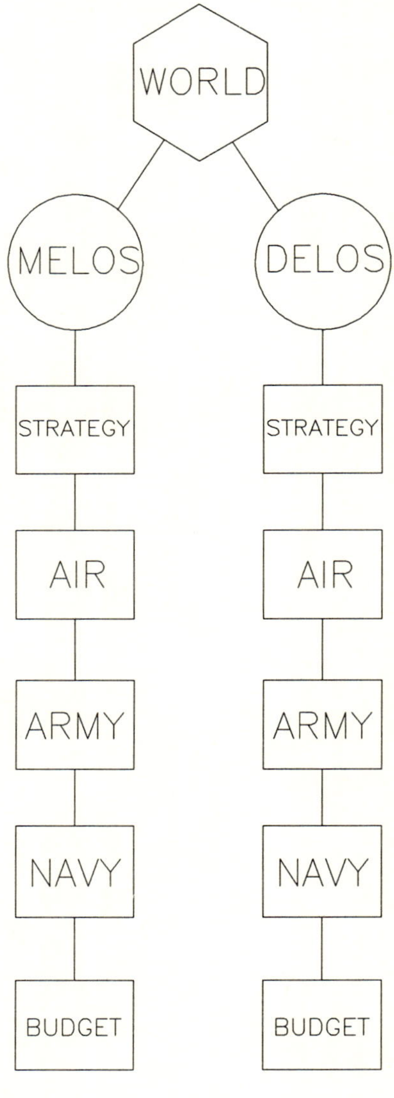

Table 4.1 ARES Functions

A. PRINCIPAL SIMULATION FUNCTIONS

TOPLEVEL – redefines IQLISP as ARES environment
GREET – prints a welcome to ARES environment
MENU – manages user interaction with ARES
DATA – controls the input of data for the run
INITIAL – sets initial conditions for run
DRIVER – controls change in simulation time
REACT – controls change in MELOS
OPERATE – controls change in DELOS
SUPPLY – allocates budget among services
GRAPH – generates a graph of run results
REPORT – generates output for screen and disk

B. UTILITY FUNCTIONS

FIND – retrieves properties
STORE – stores properties
READIN – accepts keyboard input
PAUSEN – allows N–second program suspension
APPEND – builds lists out of primitives
ATOMCAR – returns the first element of an atom
ATOMCDR – returns the remainder of an atom

n is an arbitrary number of periods, representing the number of years of simulation time; it must be a non-negative integer

X_i and Y_i represent the annual aggregate defense budget for DELOS (X) and MELOS (Y) for the i-th year; they must be positive integers

$STRATEGY_i$ may be any one of five values, representing the strategy of DELOS in the i-th year:

BATTLE - for Air/Land Battle
CONTROL - for Ocean Control
POWER - for Power Projection
INVADE - for Invasion Force
DEFENSE - for Continental Defense

This data input is constructed into a list that is referred to as the "history" of the system. The function DATA assists the user in constructing this history by prompting the user first for the number of periods in the run. In turn, for each period of simulation time, DATA prompts the user to enter values for DELOS strategy, DELOS annual aggregate military budget, and MELOS annual aggregate military budget. Valid values for these three independent variables are described above.

Now that the outlines and operation of the ARES simulation system have been described, the behavior of the system can be addressed.

4.4 Interacting with ARES

The user enters the ARES simulation environment by giving the command "IQ ARES" from the DOS system level. The ARES Arms Race Simulation is then activated. First it prints a greeting and then presents the user with a menu of options. To set up a run, the user selects options 1 and 2 to input the data and initial conditions for the current run. In doing this, the user and the program interactively create two things:

1. A bi-polar world of two nations, DELOS and MELOS: Each nation has a military strategy, and a military organization consisting of an army, a navy, and an air force of sizes and compositions given as initial conditions.

2. A script for the future history of the world, or rather, a partial script for world history, including two data types:

a. The military strategy of Delos for each year;
b. Annual aggregate military budgets for Delos and Melos for each year.

These two data items, DELOS strategy and world military expenditures, must be supplied by the user once for each year of simulation time. Once this information has been supplied to ARES, the simulation run may be performed by selecting option 3. After running, the results of the run are printed on the screen and saved in a disk file named "ARESDATA.OUT" for later analysis. We will now turn to a discussion of how the results are derived.

4.4.1 ARES Strategic Logic

ARES "action-reaction scenarios" are given in Table 4.2. The precise manner in which these scenarios are realized within ARES can be observed through an examination of the REACT function (see Appendix C for complete code listings). The core of the REACT function is a set of five statements of the following form:

```
(COND ((AND
  (GE (GET 'DELOS 'AIR)(GET 'DELOS 'NAVY))
  (GE (GET 'DELOS 'NAVY)(GET 'DELOS 'ARMY)))
    (PUT 'MELOS 'BATTLE 'STRATEGY))
```

This statement is called a "conditional" or a "production," and it is a collection of such productions (a "production system") that constitutes the control structure of ARES. This one says that if DELOS' air force is greater than or equal to DELOS' navy and DELOS' navy is greater than or equal to DELOS' army in size, then the strategy adopted by MELOS is air-land battle. In other words, if MELOS observes that the forces it faces are ordered AIR> =NAVY> =ARMY, then MELOS infers that the strategy it faces is one of power projection and that therefore a logical and prudent strategy to put in place is one of air-land battle.

It will be seen from the preceding example of ARES strategic logic that the control structure governing decision making by city-states in the system is quite elementary. MELOS' "power of observation" is limited to detecting the absolute ordering of the enemy's forces, and its strategic cunning is limited to a rigid and mechanical application of predefined responses to the various configurations of forces detected in the environment. Nonetheless, this simple control structure yields complex patterns of interactive behavior as the dynamics of the system unfold.

4.4.2 Hypothesis Testing in ARES

We can demonstrate the behavior of city-states in the ARES environment by means of an example run. We begin with a theoretical question: Why do arms races start? It is difficult to find attempts to answer this question in the literature of the Richardson Tradition because in the traditional formulation of Richardson models, interaction is represented by reaction coefficients in systems of simultaneous equations. The prospect of change in these coefficients (called diachronic change and is a change in the underlying structure of the system) is a technical problem of possibly insurmountable difficulty. Few scholars have attempted to model the process of structural change in arms races. Efforts such as Schrodt's demonstrate the technical mastery generally required to approach the problem of diachronic change using traditional methods.[10] In view of the difficulties posed by the problem of representing diachronic change, then, it is easy to see why few have ventured to wonder "how nations come to adopt their particular reaction equations,"[11] much less ask how they might change. Consequently, questions of how and why arms races start remain largely unexplored within the Richardson Tradition.

Table 4.2 ARES Scenarios

SCENARIO	DELOS ACTIONS	
	Strategy	Resource Allocations
Power/Battle	Power Projection	AIR/NAVY/ARMY
Invade/Defend	Invasion Force	NAVY/ARMY/AIR
Defend/Control	Continental Defense	ARMY/AIR/NAVY
Control/Power	Ocean Control	NAVY/AIR/ARMY
Battle/Invade	Air–Land Battle	AIR/ARMY/NAVY

SCENARIO	MELOS REACTIONS	
	Strategy	Resource Allocations
Power/Battle	Air–Land Battle	AIR/ARMY/NAVY
Invade/Defend	Continenal Defense	ARMY/AIR/NAVY
Defend/Control	Ocean Control	NAVY/AIR/ARMY
Control/Power	Power Projection	AIR/NAVY/ARMY
Battle/Invade	Invasion Force	NAVY/ARMY/AIR

An exception is Patrick Morgan, who pursues the problem of what turns a simple situation of mutual hostility into a classic arms race situation.[12] That is, within the phenomenon of armaments accumulation, how is Huntington's pre-arms race equilibrium transformed into an unstable dynamic accumulation process--an arms race.[13] Morgan posits "strategic surprise" as the catalyst of accelerated accumulation patterns that may be associated with the onset of war. Employing a variety of historical cases to illustrate the idea, Morgan suggests that when one party of a conflict dyad perceives that the other is on the verge of a qualitative transformation of force capabilities which threatens to alter the balance of forces in an unacceptable fashion, then the party perceiving this new threat may be tempted to implement a "strategic surprise" with the aim of neutralizing the incipient threat. Though Morgan does not attempt to model this process, we can explore precisely this hypothesis within the ARES simulation environment.

Let us imagine that we are in the year 490 B.C. By virtue of their guts and luck, Athens has just led the Hellenes to victory at Marathon. Yet as the first land power of the Hellenic world, it was the Spartans who had followed a strategy of continental defense, concentrating their efforts and resources on the deployment of land armies. In this same year, Themistocles the Athenian formed his strategy of sea control. Let us imagine that after Marathon the Spartans were stimulated to set in place a strategy of power projection and, further, that this state of affairs obtains for a period of twelve years, until the year 478 B.C. In this year, the confederacy agreed that Athens should succeed Sparta in the leadership role of the confederacy. The Athenians were elated, no doubt. We shall further conjecture that in their enthusiasm they altered their strategy from one of sea control to one of power projection. What are the consequences of the Athenian strategy change for patterns of armaments accumulation in the Peloponnese?

To explore this question in the ARES environment, we will introduce two simplifying assumptions. First we will assume that Athens and Sparta implemented complete bi-lateral disarmament after the battle at Marathon in 490, starting then with force postures equal to zero; this assumption is not unrealistic at least in the case of Athens, and is necessary because the data collection effort poses problems beyond the scope of this book. Second, we will assume uniform and constant defense expenditures by both city-states at the level of ten monetary units per year over the thirty-year time-series; while this assumption is stronger and its realism less defensible, it can be justified on the grounds that even Thucydides argued it was not possible to know the true levels of defense expenditures in the Peloponnesian War. Modern classicists concur: "The few figures in the surviving sources are too skimpy, too unreliable, and too often contradictory to underpin the mathematics."[14]

It may be useful to recall several things about the system world of ARES. For one thing, it is an asymmetrical bi-polar world, one in which DELOS acts and MELOS reacts. Second, and unlike most models of armaments accumulation processes, military expenditure is an independent variable. The strategy of MELOS and the structure of the military forces on either side are the "dependent" variables in ARES. Thus, what we are looking for in terms of results are changes in the pattern of armaments accumulation following from interaction among strategies and force postures of the two city-states.

Running under the assumptions outlined above, ARES yields the results shown in Table 4.3. Following the tabulated data is Figure 4.3, a graph tracing the simulated force posture timepaths over the course of the thirty-year time-series. The graph reveals an interesting finding. In the simulated Hellenic world of ARES, the growth of the DELOS air forces (DAF) eclipsed the growth of their naval forces (DN) in the simulate year 466, reflecting the power projection strategy adopted in the year 478. The decision maker in MELOS, believing that this now-visible transformation in the force posture of DELOS must reflect a long-standing set of strategic intentions--intentions that MELOS had been blissfully unaware of until now--is suddenly faced with a radical tranformation in the strategic environment. In response, MELOS immediately shifts from a strategy of power projection to air-land battle, as if to prepare for the coming combats and implement strategic surprise.

The curious aspect of this result is that in the real year 466 B.C., the Spartans did in fact change their policy, severing diplomatic relations with Athens over the issue of alleged Athenian violation of an arms control agreement prohibiting the constuction of walls. One must conclude, however, that this simulation result is a spurious one if only for the fact that it is highly unlikely the Hellenes invested anywhere near the assumed amounts on their air forces.

4.5 Problems and Prospects

A serious critique of ARES will leave an inevitable impression that, at best, the exercise leaves much to be desired. Perhaps this is inevitable because experiment and exploration were principal goals motivating the creation of ARES. The heuristic nature of these goals underscores the fact that ARES is a first generation research product. As the initial effort in what is conceived as a multiple-generation project to devise new classes of simulation systems, ARES should demonstrate the operation of the new class of formalisms in an uncluttered fashion. It is desirable, therefore, that as the first generation in the deployment of a new methodology, ARES remain as simple, technically and conceptually, as is possible. Later generations can elaborate the more sophisticated techniques availible in the AI toolbox.

Heuristic or not, however, there exist criteria against which the performance of ARES can be assessed. Tuncer Oren provides a comprehensive set of criteria for assessing the adequacy of simulation studies.[15] This section concentrates on theoretical aspects, comparing ARES with models in the reference group, the models of the Richardson Tradition (the technical review, concentrating on methods, will be found in Appendix B).

4.5.1 Assessing ARES

ARES was a response to perceived needs of the Richardson Tradition. As an attempt to press new methods into the service of the Richardson Tradition, ARES leaves ample room for improvement. This is unsurprising, for the project concerns itself not solely with methods, but also with their relationship to the process of knowledge building in a particular problem domain, that of the study of military bureaucracy. How satisfactory is ARES in meeting some of the long-standing problems of the Richardson Tradition?

Table 4.3 ARES Experimental Results

		DELOS				MELOS			
(T)	YEAR	NAVY	ARMY	AIR	STRATEGY	NAVY	ARMY	AIR	STRATEGY
0	490	0	0	0	CONTROL	0	0	0	BATTLE
1	489	5	2	3	CONTROL	3	2	5	POWER
2	488	10	4	6	CONTROL	6	4	10	POWER
3	487	15	6	9	CONTROL	9	6	15	POWER
4	486	20	8	12	CONTROL	12	8	20	POWER
5	485	25	10	15	CONTROL	15	10	25	POWER
6	484	30	12	18	CONTROL	18	12	30	POWER
7	483	35	14	21	CONTROL	21	14	35	POWER
8	482	40	16	24	CONTROL	24	16	40	POWER
9	481	45	18	27	CONTROL	27	18	45	POWER
10	480	50	20	30	CONTROL	30	20	50	POWER
11	479	55	22	33	CONTROL	33	22	55	POWER
12	478	60	24	36	CONTROL	36	24	60	POWER
13	477	63	26	41	POWER	39	26	65	POWER
14	476	66	28	46	POWER	42	28	70	POWER
15	475	69	30	51	POWER	45	30	75	POWER
16	474	72	32	56	POWER	48	32	80	POWER
17	473	75	34	61	POWER	51	34	85	POWER
18	472	78	36	66	POWER	54	36	90	POWER
19	471	81	38	71	POWER	57	38	95	POWER
20	470	84	40	76	POWER	60	40	100	POWER
21	469	87	42	81	POWER	63	42	105	POWER
22	468	90	44	86	POWER	66	44	110	POWER
23	467	93	46	91	POWER	69	46	115	POWER
24	466	96	48	96	POWER	71	49	120	BATTLE
25	465	99	50	101	POWER	73	52	125	BATTLE
26	464	102	52	106	POWER	75	55	130	BATTLE
27	463	105	54	111	POWER	77	58	135	BATTLE
28	462	108	56	116	POWER	79	61	140	BATTLE
29	461	111	58	121	POWER	81	64	145	BATTLE
30	460	114	60	126	POWER	83	67	150	BATTLE

Figure 4.3 Simulated Force Postures

at Outbreak of the Peloponnesian War

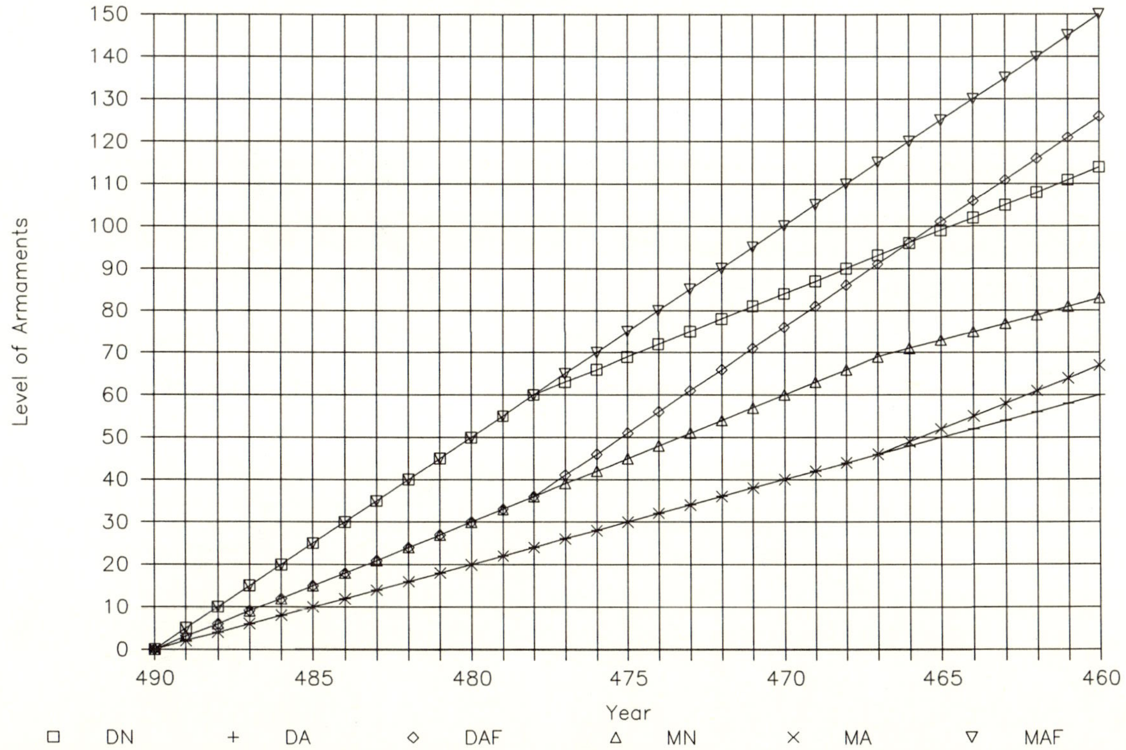

ARES demonstrates the knowledge representation capacity of qualitative languages as against the traditional quantitative modes of discourse. As suggested in Chapter Three, the qualitative languages can support any formalism supported in quantitative languages, as well as others that would be either impracticable or impossible in quantitative languages. Using the class of formalisms to which ARES belongs, it is possible to expand our conception of the state spaces within which armaments accumulation processes occur. As this seems to be a necessary condition for progress in the study of armaments accumulation processes, this aspect of ARES is counted as a success.

As Intriligator argues elsewhere, a "criticism of all the models discussed in Sections 1 and 2 is that they treat an arms race from the outside as a mechanistic model, rather than from the inside in terms of decisions of defense planners."[16] Intriligator moved toward solving this problem, though what he refers to as "grand strategy" would in a conventional conceptual framework be called "operational strategy." Thomas Saaty also suggests that a "more general theory is needed that would be less mechanical and would consider rational behavior accounting for the relationship between arms buildup and the occurrence of conflicts."[17] ARES goes a certain distance toward meeting this traditional criticism of Richardson-type models. ARES is still mechanical in that it is a deterministic system with no stochastic component. ARES is also mechanical in the technical sense that it is an "automaton," a finite state machine, by definition. There is, however, a rudimentary decision algorithm--a mechanism for choice--at the heart of the model. Thus, ARES is not mechanistic in the sense of relying on physical conceptions of the relations of matter and space alone. The ARES decision logics introduce meaning, purpose, intentionality, and goals in the form of strategic plans. This is seen as a significant contribution to the study of arms accumulation.

Still, ARES is often criticized as too simplistic. The decision logic implemented in ARES uses the most elemental of control structures available in artificial intelligence research, that is, the first-order production system of Alan Newell and Herbert Simon. This is vintage early 1960s technology. In defense of this model, it is important to note that ARES is purposely designed to be extremely simple and that the use of these elemental techniques is not accidental. I view this approach as appropriate in introducing this relatively unknown methodology to the discipline, as a step in the transition of the Richardson Tradition to the use of modern mathematics. More complicated, rule-based simulation work is required, and in fact it is in progress now by Mefford and Hybel, Sylvan and Majeski, Sylvan, Daniel Bobrow, Brian Ripley, Margee Ensign, Schrodt, Etcheson, and others. With all of this more sophisticated material now hitting or soon to be hitting the professional journals, an elementary primer is called for and needs to be presented in a familiar idiom and pitched in the context of the concerns of the tradition.

So ARES is tantalizing but unsatisfying. The example run shows how such a simulation environment can be used to explore factual and/or counter-factual hypotheses about the interaction of strategy and bureaucracy in internation armaments accumulation processes. More than anything, however, ARES is heuristic and experimental. In order to examine fully the potential utility of AI control structures in this problem domain, much more extensive analyses will be required. This suggests a series of models.

The next generation of simulation environments in project MONADYSIM will incorporate a more advanced control structure called a second-order production system. This meta-rule approach will permit the model to deal with a wider variety of environmental transformations and achieve a greater degree of structural dynamism. This may include such features as look-ahead and sector-specific sensitivity to rates of change in weapons stocks. It should also incorporate more sophisticated search algorithms and more complex data structures for precedent-based decision making.

The third generation of MONADYSIM models remains only in foggy outline at present. It would endeavor to employ state of the art AI technology. This could allow the simulated strategic decision maker to move beyond precedent-based decision rules to an endogenous capacity to generate novel strategic action plans on the fly. Such an approach would open the way for simulation of true diachronic transformation in armaments accumulation processes. This is precisely what is required in order to elaborate emancipatory strategies for the understanding of armaments accumulation phenomena. ARES constitutes another halting step in this direction.

4.6 Summary and Conclusions on Computational Model

This chapter has reported an experimental artificial intelligence-inspired simulation system named ARES. The system was designed to support the study of strategic interaction between two city-states. The presentation generates several conclusions:

§ *The problems of modeling ancient conflict turn out to be precisely the same ones associated with the modeling of contemporary conflict.*

§ *The first-order production system control structure used in ARES allows the representation of detailed strategic logics.*

§ *Although ARES is only experimental, it provides a starting point for discussing the possibilities of AI in the modeling of armaments accumulation.*

§ *There are inherent costs and limitations associated with an AI approach, but these are not so great as to prohibit further explorations of the method.*

§ *ARES points down a path of new possibilities for exploiting AI techniques in the modeling of armaments accumulation processes, though it takes only a first and tentative step down that way.*

Subsequent steps are explored in the next chapter, which compares ARES with traditional structural approaches in the Richardson Tradition, showing the limits of each.

Notes

1. As an anonymous reviewer of this chapter perceptively noted, "This piece has absolutely nothing to do with Hellenic Greece!!" The ARES model represents a **simulate** environment, not a real one. Readers should be cautioned not to mistake the arbitrary labels used in ARES for an empirical or historical reference; the purposes here are rather theoretical and methodological.

2. Heroditus, The Histories of Heroditus, trans. George Rawlinson (Chicago: Encyclopaedia Britannica, 1952), Book 8, pp. 143-144.

3. Plutarch, Lives of Noble Grecians and Romans, trans. John Dryden (Chicago: Encyclopaedia Britannica, 1952); see "Themistocles."

4. Thucydides, The Peloponnesian War, trans. F. Crawley (New York: Random House, 1951), Book 1, p. 23.

5. Lewis Fry Richardson, Arms and Insecurity: A Mathematical Study of the Causes and Origins of War (Pittsburgh: Boxwood Press, 1960).

6. Albert Wohlstetter, "Is There a Strategic Arms Race?" Foreign Policy 15 (Summer 1974), p. 8.

7. Samuel P. Huntington, "Arms Races: Prerequisites and Results," Public Policy 8 (1958), p. 50.

8. Colin S. Gray, "The Arms Race Phenomenon," World Politics 24 (October 1971), p. 75.

9. Frank Moulaert, "On the Nature and Scope of Complex Conflicts," Peace Science Society (International) Papers 29 (1979-1980), pp. 48-66.

10. Marwin Simaan and Jose B. Cruz, Jr., "A Multistage Game Formulation of Arms Race and Control and Its Relationship to Richardson's Model," in Modeling and Simulation, ed. W.G. Vogt and M.H. Mickle (Pittsburgh: Instrument Society of America, 1973), Vol. 4, pp. 149-153; and also their "Formulation of Richardson's Model of Arms Race from a Differential Game Viewpoint," Review of Economic Studies 42 (1975), pp. 67-77; and Philip A. Schrodt's more elaborate effort, "The Effects of a Richardson Arms Race on the Distribution of Power in an International System" (Ph.D. dissertation, Indiana University, 1976).

11. John V. Gillespie, Dina A. Zinnes, G.S. Tahim, Philip A. Schrodt, and R.M. Rubison, "An Optimal Control Model of Arms Races," American Political Science Review 71 (1977), p. 226.

12. Patrick M. Morgan, "Arms Races and Strategic Surprise," paper presented to the annual meeting of the International Studies Association, Los Angeles, March 19-22, 1980; see also Martin Patchen's look at the movement from arms race to war, "When Do Arms Buildups Lead to Deterrence and When to War?" paper presented to the second World Peace Science Congress, Rotterdam, June 8, 1984.

13. Huntington, "Arms Races: Prerequisites and Results," op. cit.

14. M.I. Finley, Economy and Society in Ancient Greece (New York: Viking Press, 1953), p. 48.

15. Tuncer I. Oren, "Concepts and Criteria to Assess the Acceptability of Simulation Studies: A Frame of Reference," Communications of the ACM 24 (1981), pp. 180-189.

16. Michael D. Intriligator and Dagobert L. Brito, "Formal Models of Arms Races," Journal of Peace Science 2 (1976), p. 82.

17. Thomas L. Saaty, Mathematical Models of Arms Control and Disarmament: Applications of Mathematical Structures in Politics (New York: Wiley and Sons, 1968), p. 50.

5 STRATEGIES OF ARMS ACCUMULATION RESEARCH

The goal of this chapter is to criticize the ARES model. In the process, we will explore methods for formalizing the concept of diachronic change in armaments accumulation processes. Conceptualized as a network of planning behaviors, strategic interaction in armaments accumulation processes can be analyzed for meta-level properties associated with system transformation. Though far from exhaustively developed at present, it is possible to outline plausible research strategies for the design and implementation of conceptually adequate models of structural change in armaments accumulation processes. It is to be emphasized that ARES fails the conceptual adequacy test proposed here.

The plan of this chapter is to show, first, that representation of structural change is necessary in a conceptually adequate model of arms accumulation and second, that existing efforts contain no such representations. We begin by elaborating a conceptual framework for considering strategic behavior. Employing this framework, we will then examine the structural and adaptive approaches to modeling arms accumulation behavior. This will demonstrate that we remain far from realizing a conceptually adequate model of arms accumulation processes.

5.1 Planning Arms Accumulation

". . . the inner connexion of the whole act of War . . . [is] the plan for a War."

Vom Kriege, Bk 3, Ch 1

5.1.1 Purposive Agents in the Acquisition of Capabilities

Loosely defined, strategy refers to a plan to achieve a goal within some context or environment. Strategy by definition must deal both with what one proposes to achieve and how one proposes to achieve it. When considering the concept of strategy in the context of military planning for arms accumulation, however, the matter grows complex. This is due to the necessity of considering the structure of goals at different levels in social organizations. The resulting complexity can be illustrated by considering the goal structures at different levels for a concrete case.

In 1909, Great Britain constructed eight capital warships (i.e., battleships or battle cruisers). From the perspective of the managers of the shipyards and the Admiralty's procurement office, the goal was simply to build eight ships of a particular type. Ship production was the primary goal sought at this level in the social organization. Such a goal involves many sub-goals, such as deciding what class of ships to build, exactly what sort of guns it should have and where they should be built, how much armor in which locations, etc. The scope of the goal structure at this level in the organization is bounded by the mission assigned to the navy.

If one moves up one level of social organization, however, the structure of goals changes. From the perspective of the British General Staff, building eight ships ceases to be a primary goal and becomes a means toward a higher level goal. The higher level or primary goal here may consist in the desire to maintain a "two-power standard," that is, to maintain naval superiority over any possible combination of two rival powers, say France and Germany. The eight ships are, for the General Staff, a sub-goal. Other sub-goals at this level of the organization may include maintaining additional types of vessels beyond the capital classes, as well as an army of a given constitution and distribution, various foreign bases of operations, and so on. Thus, the scope of the goal structure at the general staff level of the organization is bounded by the local primary goal and conditioned by the strategy of the military as a whole.

Moving up the organizational ladder one more level to the level of the top political leadership, what is the goal of the British General Staff again becomes a sub-goal for the Prime Minister. The goal of the Prime Minister may consist in the maintenance of global geo-political hegemony in a fashion consistent with the demands of important domestic constituencies. Here questions of the details of the military balance are just that--details--to be considered along with economic, political, and all the other kinds of details that the Prime Minister must weigh. The goal structure at this level, then, is bound by the global political ends of the national power structure as a whole.

This cross-section of goals pertaining to British armaments accumulation behavior in 1909 illustrates the complexity of considering armaments accumulation from a goal-oriented perspective, but the matter becomes more complicated yet if we also consider a longitudinal view. British naval history constitutes considerably more than what happened in 1909. The four hundred year period from the destruction of the great Spanish Armada in 1588 to the battle of the Malvinas/Falklands in 1982 evidenced many changes in the scope of the goals sought by British leaders at each of the three levels in the conceptual hierarchy of strategic behavior. The kinds of goals being sought in 1588 by Henry III and Francis Drake, in 1910 by Edward Grey and Winston Churchill, and in 1982 by Margaret Thatcher, occurred in completely different contexts, at the dawning, the zenith, and the twilight of empire.

A third layer of complication is added to the cross-sectional and longitudinal complexities: There exist "Others" in the environment. Goal-seeking agents at all levels in the social organization recognize that one of the most difficult challenges is the existence of intelligent Others. These Others in the environment may actively attempt to block the achievement of various goals or may have plans conflicting with those at one or more levels of the organization. This implies that plans and goals may be in conflict. Conversely, Others may catalyze or cooperatively assist in actualization of the intentions and values expressed in ones own plans. These interactions must be explicitly taken into account in the creation of plans.

This illustration is abstracted in Figure 5.1. The remainder of this section is devoted to a discussion of this conceptual model (national command authority--NCA; ministry of defense--MOD; autonomous military services--ARM; and rules and meta-rules--R, M). This will help to relate the goals at different organizational levels to the problem of conceptual adequacy. For the time being, it is sufficient to distinguish among "grand strategy," "military strategy," and "operational strategy" (or simply, "operations"), and to note that all three types of strategy turn on the concept of "planning."

5.1.2 A Conceptual Perspective on Strategic Behavior

A "strategy" is a social action plan. A plan, generally speaking, is any hierarchical process that can control the order in which a sequence of actions is performed. This concept of a plan follows G.A. Miller, E. Galanter, and K.H. Pribram,[1] though our use is more restricted in that we are specifically concerned with social action. We will focus on three particular varieties of social action plans or strategies: political, military, and administrative action plans, which will be referred to as "grand strategy," "military strategy," and "operational strategy," respectively.

Grand strategy is the province of political leaders. It refers to plans about the relationship of political ends to the ways and means of policy. The means of policy include any and all resources the policy maker can bring to bear in problem solving: political, economic, moral, military, psychological, diplomatic, and so on. This definition follows conventional usage. Unfortunately, as Liddell-Hart has observed, the "sorry state of peace, for both sides, that has followed most wars can be traced to the fact that, unlike [military] strategy, the realm of grand strategy is for the most part terra incognita--still awaiting exploration and understanding."[2]

Figure 5.1

Conceptual Structure of Strategic Interaction

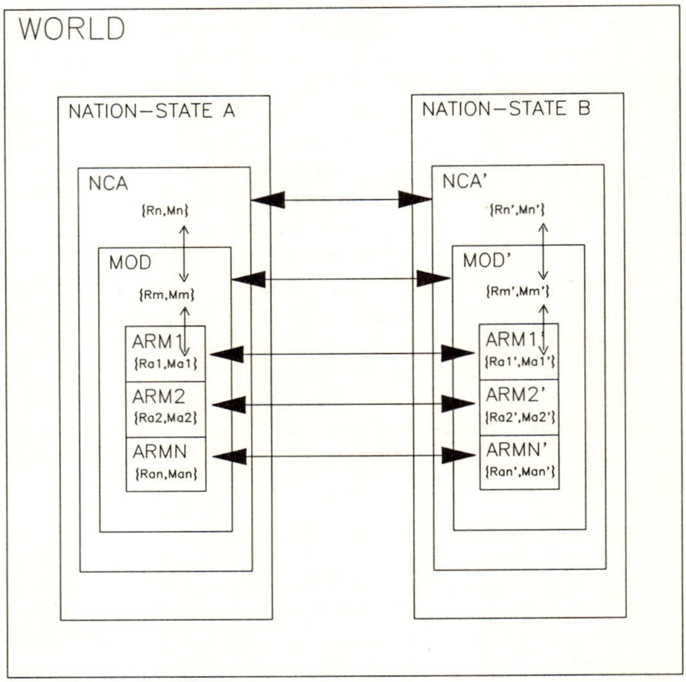

Arrows pointing down transmit goals.
Arrows pointing up transmit capabilities.
Horizontal arrows transmit environmental data.
{Rx,Mx} indicate rules and meta–rules for Unit x.

In contrast to grand strategy, military strategy is the province of military leaders, restricted in scope to the conduct or contemplation of war. There are many definitions. The great Prussian strategist Karl Von Clausewitz defined it as "the employment of the battle as the means towards the attainment of the object of the War. . . . In other words, Strategy forms the plan of the War."[3] Von Clausewitz' emphasis on the decisive nature of combat, however, allowed some of his readers to miss his central point, that is, as Abba Eban reminds us, the "object of war is not military--the destruction of the enemy--but political--the attainment of a result. . . ."[4] Recognizing this, the German General Helmuth von Moltke (the elder) highlighted constraints on both ends and means in war when he refered to military strategy as "the practical adaptation of the means placed at a general's disposal to the attainment of the object in view." Liddell-Hart refines this to read, "the art of distributing and applying military means to fulfill the ends of policy,"[5] thus making explicit the connection and subordination of military strategy to grand strategy.

As military strategy is connnected to grand strategy, so is the concept of operations connnected to military strategy. Operations refers to the administration or management of military forces in war and peace. As used here, operations includes, but is not restricted to, the ideas of tactics and logistics, terms normally reserved for the deployment of forces in combats.

The conception of strategy outlined here rests upon the notion that military strategy is a plan embedded in a network of other plans. To consider "the" plan only from the level of operations is clearly inadequate. Similarly, considering events only from the perspective of military strategy will distort analysis; the military plan alone is meaningless. Even grand strategy cannot be considered in isolation, for it obviously depends upon what happens at the other levels of conceptualization. Thus, even an individual analysis of each level will not be sufficient, for each of the three levels of analysis are intimately interactive and hence interdependent, though nominally autonomous.

Howard Tamashiro's discussion of the Egyptian decision to attack Israel in the 1973 Yom Kippur War illustrates clearly the importance of identifying and discriminating among the planning behaviors at different conceptual levels in the analysis of arms accumulation behavior.[6] The issue in this case concerns Egypt's acquisition of tens of thousands of Sagger anti-tank missiles for use against an opposing Israeli tank force numbering some 1,500. The question is, to what extent can one consider the Egyptian plan for the Yom Kippur War to have been "correct"? It turns out that the answer to this question depends entirely upon the structure of goals in the organization and precisely which set of goals one is considering:

> From a tactical perspective, the Sagger-armed Egyptian infantry went into battle largely unsupported by planes or tanks of their own. Hence, the Israelis could use artillery fire to decimate the unprotected Egyptian infantry. . . . Hence, any analysis pitting Egyptian infantry armed with Saggers against Israeli tanks supported by artillery would, in this tactical context, conclude that the Egyptians planned "incorrectly." From an operational perspective, the Egyptian army lacked the skilled manpower to wage mobile war, coordinating tanks, artillery, and aircraft in support of infantry. Saggers, on the other hand, were plentiful and easy

> to use. Large numbers of relatively uneducated troops could be trained to use the antitank missiles effectively. Hence, in this context, the Egyptian decision to rely on missile tactics was "correct." From a strategic perspective, the unsupported Egyptian Sagger teams, no matter how plentiful, could not in the long run carry the war against Israeli mobile operations. Hence, in this context, the Egyptian decision to start such an unwinnable war was "incorrect." From a political perspective, the Egyptians did not need to win the war militarily. Only an opening tactical victory . . . [was required] to raise Egyptian morale enough to make future Israeli-Egyptian negotiations feasible. Hence, in this context, the Egyptian "Sagger" decision was "correct."

Any analysis of the Yom Kippur War that proceeds solely on the basis of tactical considerations and/or military strategy, then, would find Egyptian decision making fatally flawed. Military strategy does not encompass the concept of a carefully planned defeat of one's own forces--that's just not the way planning is done in military strategy. In fact, however, it is such an analysis that is flawed, failing to take into account those elements of strategic planning we call grand strategy.

5.1.3 Requirements for a Conceptually Adequate Model

The requirements for a conceptually adequate model of armaments accumulation processes follow readily from the foregoing discussion. Based upon the conceptual framework presented as Figure 5.1, one can identify five elements necessary to the adequate model.

(1) The model must represent all three levels of organization in the state apparatus; they are logically necessary for coherence in tracing the evolution of goal-seeking behaviors. Ideally, the model should also contain a fourth level, representing the non-executive elements of the nation-state, and a fifth level, representing transnational and other non-nation-state entities in the world.

(2) The model must represent the interactions among the three levels in the executive organization of the state in order to explain the flow of changing goal structures.

(3) The model must represent the interactions at each level of the organization with purposive, intelligent others in the environment; this is the source of system dynamics.

(4) The model must represent rules of behavior for each of the levels in the organization, specifying the performances to be enacted in any given situation.

(5) The model must represent meta-rules to describe the transformation of decision rules over time; organizations survive if and only if they are able to adapt to changing situations by changing their own rules of behavior whenever the "rules of the game" change.

The requirements for a conceptually adequate model outlined above do not define an ideal model. They represent minimum requirements for a conceptually adequate model of armaments accumulation behavior. Let us examine a typical model of armaments accumulation to see how far we have progressed toward an adequate model of this phenomenon.

5.2 Structural Models of Arms Accumulation

This section reports on a typical structural model of arms accumulation behavior, assessing the conceptual adequacy of that model. It is useful to begin by setting the historical context within which the model operates.

5.2.1 Pre-World War I Arms Accumulation Planning

Napoleon Bonaparte changed war. "From the dawn of recorded history until shortly before 1800 . . . technology and organization placed limits on the abilities of armies to gather, transmit, and process information and, consequently, on their size and the nature of the missions they could perform. . . [i.e.,] on what strategy could and could not do."[7]

Napoleon Bonaparte's revolutionary armies were finally defeated in 1815, but his example revolutionized the face of war. His superb use of dispersion and concentration at the level of operations stunned his opponents and did much to enshrine his status as a military genius, but these were not novel ideas; his enduring contributions to the ways of war were organizational. Attempts to imitate the form of Napoleon's reputed military genius led to the creation and expansion of general staff organizations as permanent peacetime agencies for the first time in history. War and peace haven't been the same since.

Strictly speaking, as is shown in the British and Prussian examples, the roots of the modern general staff system predate Napoleon. The British Admiralty provides an example of a long-standing peacetime military planning agency. But historians trace the idea of a general staff, as opposed to a staff for an individual service, to the Prussians. At the beginning of the nineteenth century, the old quartermaster staff's logistical functions began to evolve into today's modern, technical bureaucratic-adminstrative apparatus. The general staff system for the first time provided a permanent general military planning agency between the sovereign and his sword, completing the linkage running from grand strategy through military strategy to operational strategy.

General or not, however, the Prussian's were never at home on the sea. Prussia's dominating landed aristocracy had always conceived of itself as a land power, first and foremost. Consequently, they never had a navy to speak of. As late as 1865, the Prussian parliament rejected a proposal to embark on a ten-year naval construction program aimed at building a "second rate" navy.[8] Two years later a similar program was finally adopted, and though it symbolized the cementing of an enduring political alliance between the agrarian Junkers and the industrial Bavarians, the construction plans were interrupted by the Franco-German War.[9] As late as 1890, the German admiralty (*Reichsmarineamt*) still had its mission defined in terms of defending the north coast and planning for cruiser attacks on enemy commerce in the event of war. There was no plan as yet for actually engaging a serious naval power. Not until the 1880s did German leaders even begin to think in terms more than continental in scope. Ever since the time of the Corsican, the "German strategy" had simply been to maintain the strongest army in Europe.

In the 1890s the German strategy changed. In a word, *weltpolitik* was born, and many German leaders became almost obsessed with the idea that, for the "general purposes of her greatness," Germany must have a first-rate navy. Kaiser Wilhelm II was on board, arguing in 1896 that "once again we can see how wrong-headed we were to begin a colonial policy ten years ago without a fleet."[10] His close aide, Admiral Alfred von Tirpitz--secretary of the navy between 1898 and 1916--tried to relate the navy to traditional Prussian concerns: "only a fleet which represented alliance value to other Great Powers, in other words, a competent battle fleet, could put into the hands of our diplomats the tools which, if used to good purpose, could supplement our power on the continent."[11] This change in German strategy sparked Anglo-German antagonism and ignited the Anglo-German naval race of 1905-1916. As Winston Churchill wrote, "The determination of the greatest military power on the continent to become at the same time the second naval power was an event of the first magnitude in world affairs. It would, if carried into full effect, undoubtedly reproduce those situations which at previous levels in history had proved of such awful significance to the Islanders of Britain."[12]

5.2.2 Lambelet's Model of the Dreadnought Race

In a series of articles, Jean-Christian Lambelet presents one of the most rigorous treatments of an arms race in the literature of the Richardson Tradition.[13] Lambelet took as his case study the Anglo-German "Dreadnought Race" of 1905-1916. Lambelet shows that Germany's naval supply behavior can be modeled as if, though it may have been interacting with Britain in some way, it was not reacting directly to changes in British shipbuilding behavior in the sense of the classic tit-for-tat, action-reaction arms race image. Lambelet gives Germany's naval supply function as:

IF (T.EQ.1905) G = 0 [5.1]

IF (T.EQ.1906) G = 2 [5.2]

IF (T.EQ.1907) G = 3 [5.3]

IF (RCS.LE.0.500) G = 4 [5.4]

G = 3 [5.5]

where

T is time
EQ is equal to
LE is less than or equal to
G is number of German capital ships produced in year t
RCS is ratio of completed ships at t-1 (Germany/Britain)

Thus, the values are exogenous for the first three years by declaration in equations 5.1 to 5.3. Specifications of the endogenous model for the period (1908-1916) are, simply, if the ratio of German capital ships to British capital ships completed in year t-1 is less than or equal to one-half, Germany builds four ships in year t (5.4); otherwise Germany builds three (5.5).

British naval supply behavior is more complex:

IF (T.EQ.1905) B = 4 [5.6]

IF (T.EQ.1906) B = 3 [5.7]

IF (T.EQ.1907) B = 3 [5.8]

$$B = 3.0*G^{L1} + 1.5*GSB^{L2} - BSB^{L1} \qquad [5.9]$$

where

T and EQ are the same as above
B is number of British capital ships built in year t
G^{L1} is the number of German capital ships built in t-1
GSB^{L2} is the sum of German ships built up to year t-2 inclusive
BSB^{L1} is the sum of British ships built up to year t-1 inclusive

As with its counterpart, the first three equations of the British sub-model (5.6-5.8) define exogenous initial conditions. The British naval supply function for the years 1908-1916 is given by Equation 5.9. This equation is very similar in form to familiar basic Richardson action-reaction model, though stated in the atypical discrete form of an economic stock adjustment function. This function specifies that British shipbuilding in year t is a positive function of Germany's capital shipbuilding program up to last year, and a negative function of the stock of British ships. Very simply put, the British are modeled as reacting directly in proportion to the size of the German high seas fleet, prudently discounting that enormous perceived threat by keeping in mind the preponderance of their grand fleet.

5.2.3 Analysis of Structural Approach

Lambelet's model has much to recommend it in comparison with other structural models in the Richardson Tradition. In contrast to many models of arms accumulation, Lambelet's model focuses on discrete variables (particular categories of weapons), rather than the usual amorphous budget variable. The asymmetry of control structures for Britain and Germany reflects the firm empirical grounding of Lambelet's effort, comparing favorably with the more common "mirror image" control structures. The Dreadnought Race model performs well within the temporal confines for which it was designed, with good agreement between the outputs of the model and the historical data (R = .78 for Britain and R = .90 for Germany). Overall, his effort must be regarded as among the most carefully crafted pieces of scholarship in the Richardson Tradition.

In terms of the conceptual structure of strategic interaction presented earlier, however, the Dreadnought Race model is judged somewhat more severely. What Lambelet has produced is a state representation of only the inner-most level of the three level interaction hierarchy. His is a model of the supply function for one arm of the military forces. It neglects the other arm of the military (the army), as well as the remainder of the associated apparatus. In the lexicon of Figure 5.1, Lambelet represents only the inner-most set of rules, but none of the higher-level rules and none of the meta-rules. In other words, the Dreadnought Race model deals only with the rules of operational strategy, neglecting military strategy, grand strategy, the relations among the different strategic levels, and changes in those relations. Consequently, it is historically bound (i.e., useful only within the tightly bound historical context of 1905-1916). According to the criteria adopted at the beginning of this chapter, then, one must conclude this model is not conceptually adequate. This conclusion is illustrated in Figure 5.2.

In summary, while Lambelet's Dreadnought Race model is among the finest of its class, it is far from being a conceptually adequate model of armaments accumulation processes. Lambelet's reduction of all of strategy to simple operations is unfortunately typical of the Richardson Tradition. It is clear that assertions such as "the problem of *grand strategy* is that of choosing both a rate of fire and a targeting strategy"[14] are conceptually confused, mistaking operational strategy for grand strategy while neglecting both military strategy and grand strategy. The ultimate consequence of failure to represent higher level strategic behavior in models of arms accumulation is echoed in Von Clausewitz' famous comment: "War can never be separated from political intercourse, and if . . . this is done in any way, all the threads of the different relations are, to a certain extent, broken, and we have before us a senseless thing without an object."[15]

5.3 Representing Change in Models of Arms Accumulation

This section will report an attempt to deal with the conceptual shortcomings of structural models of arms accumulation. In the process, a more rigorous definition of the idea of change will be developed. It will then be seen that these recent efforts fail to satisfy our criterion of conceptual adequacy. What ARES leaves unanswered are some key questions about the generation of plans for arms accumulation.

5.3.1 Change in Arms Accumulation Trajectories

The coming of war in 1914 transformed patterns of armaments accumulation in the German military. Entrenched in the officer cadre of the German army and the upper house of the Reichstag, the conservative Prussian aristocracy gained greater weight in the deliberations of state with war. The German navy's share of the defense Reichsmark had grown from less than 10% in 1890 to a full third of the total in 1910, but with German soldiers falling daily on the battlefield, the Prussians were able to reassert the traditional budget priorities.[16] The new "landships" (tanks) helped divert

Figure 5.2
Conceptual Structure of Structural Model

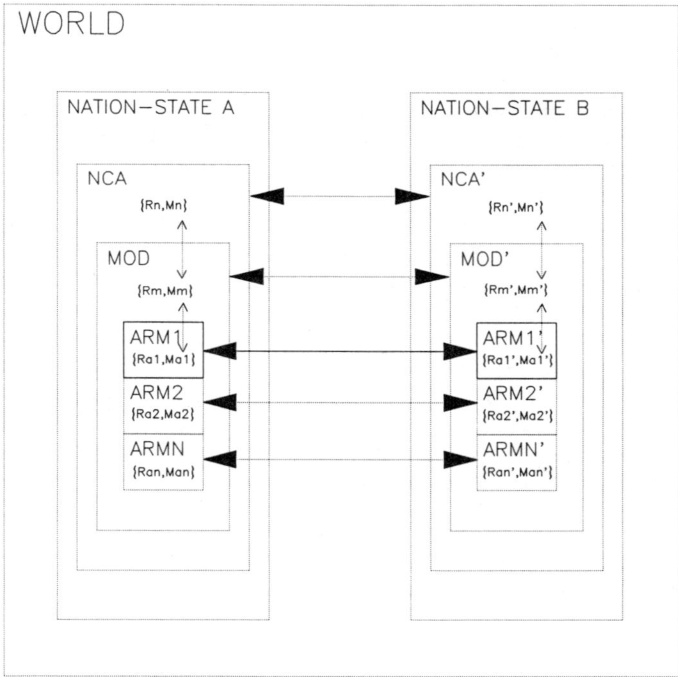

Arrows pointing down transmit goals.
Arrows pointing up transmit capabilities.
Horizontal arrows transmit environmental data.
{Rx,Mx} indicate rules and meta–rules for Unit x.
Dashed lines indicate concepts not included in this model.

attention from plans for more battleships. Interservice rivalry thus cut into the navy's resources. As the Kaiser became increasingly enamored with the new submarine technology, budget priorities changed even within the German navy. Eventually the ambitious German capital shipbuilding program ground to a halt: "Starting in 1917 work on these ships was pursued only to give some work to the naval yards; it was almost [*sic*] completely stopped after November 18."[17] This was the mundane end of the Anglo-German Dreadnought Race.

Lambelet's model deals neither with the beginning nor the end of the Dreadnought Race. This is odd, for one of the most interesting things about the Anglo-German naval armaments competition of 1905-1916 is the way it ended. Some analysts have uncritically assumed that arms races inevitably cause and/or result in war. A peculiarity of the Dreadnought Race is that precisely the converse seems to have happened: A war in part caused an arms race to end. The high resource demands of seemingly interminable trench warfare on the ground in Europe deprived the British and German naval establishments of the resources required to carry on in the capital ship construction competition. Evolving military strategy and operations through the war contributed as well. Unfortunately, these issues were beyond the scope of Lambelet's effort, and the Dreadnought Race model does not include the rules needed to represent these changes. Indeed, a large and unwieldy chunk of FORTRAN code would be required to model the changes in strategy and resource allocations exhibited in German and British armaments accumulation patterns during World War I. This would clutter Lambelet's elegant and parsimonious model considerably.

It will nonetheless serve our didactic purposes to develop Lambelet's model along these lines. There presumably exists, then, an additional set of rules--as yet unmodeled--to describe the supply behavior of the German army. Legislation to expand the German army was promulgated in 1905 and again in 1912. Suppose for the sake of discussion that upon mobilization for war the German army increased in size by a factor of three. Following Lambelet's notation, this would give the following rules:

IF (T.LT.1912) G' = 0.70X [5.10]

IF (T.GE.1914) G' = 3.00X [5.11]

G' = X [5.12]

where

G' is the number of German army divisions
LT is less than
GE is greater than or equal to

These German army supply rules should be good for the period 1905 to 1917 or so, subject to empirical determination of the value of X and perhaps some adjustments of the parameters preceding X in Equations 5.10 and 5.11.

At the same time that resources available to the German army expand, those available to the German navy contract. Thus, there must exist an additional set of rules for the supply behavior of the German navy after 1916. Taking up where Lambelet's model leaves off:

IF (T.EQ.1917) G = 1 [5.13]

G = 0.500 [5.14]

Rules such as Equation 5.13 (and by implication the initializing rules in the Dreadnought Race model, 5.1-5.3 and 5.6-5.8) express no functional relation and thus may be thought of as "degenerate rules." They do serve to display quite clearly the bald *ex post facto* nature of all such models in the Richardson Tradition. Whether rules 5.10 through 5.14 are the best possible representations for the behavior of the German military during this period is an empirical question that need not concern us here. The important point is that some such rules must exist, and these particular ones will do for present purposes. Given the existence of such rules, however, certain questions naturally assert themselves: Where do these rules come from? If they represent plans to achieve goals, how are the plans generated? The answer, of course, is that there is a goal-seeking agent at a higher level than the army or navy. This higher agent is able to alter the pattern of resource allocations to the various military services in support of a changing situation.

A model of the armaments accumulation processes surrounding World War I, therefore, has to be "intelligent" enough to know when it is appropriate to replace the German naval supply function in Rules 5.6-5.9 with Rules 5.13-5.14 and so on. In most models of armaments accumulation, Lambelet's included, questions about the sources of rules and change in planning are not dealt with. It seems clear that such questions are central to the design and production of a conceptually adequate model of armaments accumulation processes.

5.4 Finite State Machine Models of Adaptive Arms Planning

In the previous chapter (and in other papers) I have presented a model of strategic planning in the Peloponnesian War that attempts to come to terms with some of these issues.[18] The ARES model contains a rule structure representing this type of change in the internal resource allocations among military services of a nation-state. The control structure in ARES consists of a set of rules that specifies alternative sets of resource allocation priorities. Depending upon perceptions as to the state of the environment (defined in terms of military deployments of Others), an appropriate rule is selected for implementation.

To elaborate on an ARES-type model for World War I armaments accumulation processes, we can define two military strategies for Germany, "land-battle" and "power-projection." For Germany to adopt a military strategy of land-battle means that they will distribute resources to their military services in a ratio of nine to one in favor of the army over the navy. If, on the other hand, Germany adopts a strategy of power-projection, the ratio becomes six to one, improving the relative position of the navy. Whether they adopt one strategy or another depends on interactions with Others in the environment. Based on the order of battle of Others, the decision-making mechanism in an ARES model makes inferences about the strategy of the Other. Decision makers cannot see the plans and intentions of Others, so they act on the capabilities they can see. We will say that when faced with an opponent whose order of battle indicates a strategy called "sea-control," the Germans will use the rule for power-projection to guide their own planning behavior. However, if they infer that the Other's strategy is land-battle, the Germans will respond in kind with land-battle. This yields two rules for German military strategy at the MOD level:

IF LAND-BATTLE, THEN (G = 9, G' = 1) [5.15]

IF SEA-CONTROL, THEN (G = 6, G' = 1) [5.16]

Using control structures of this type, one is able to model military strategy at the MOD level of the organization. The conceptual structure of such a model is shown in Figure 5.3. Although it has not been implemented, this approach also permits modeling the rules of grand strategic behavior at the NCA level of the organization.

5.4.1 Analysis of Finite State Machine Models

Finite state machine models represent a significant advance in the art in that they allow the modeling of the complex set of relationships among goals at different levels of social organization in armaments accumulation processes. This brings our models of arms accumulation closer to the criterion of conceptual adequacy set forth at the outset of this chapter. However, as we shall argue, this does not quite do it.

Several recent international relations modeling efforts have employed the mathematical formalism known as production systems, a type of control structure sometimes classified as a "generator."[19] These models are automatons. Thorson et al. develop models of foreign policy decision-making inference behavior using an automaton.[20] As reported above, I also developed a model of strategic policy decision-making inference behavior using an automaton. These two approaches differ in significant ways, but they are similar in the sense that the two different types of automatons belong to the same class of finite state machines. Formally, the class of automatons used in these efforts is defined as a six-tuple,

$$\{Q, A, \Omega, Z, \phi, \beta\}$$

where

Q is the state space
A is the input syntax
Ω is the output syntax
Z is the transition function
ϕ is the output function
β is the initial state

Figure 5.3

Conceptual Structure of Adaptive Model

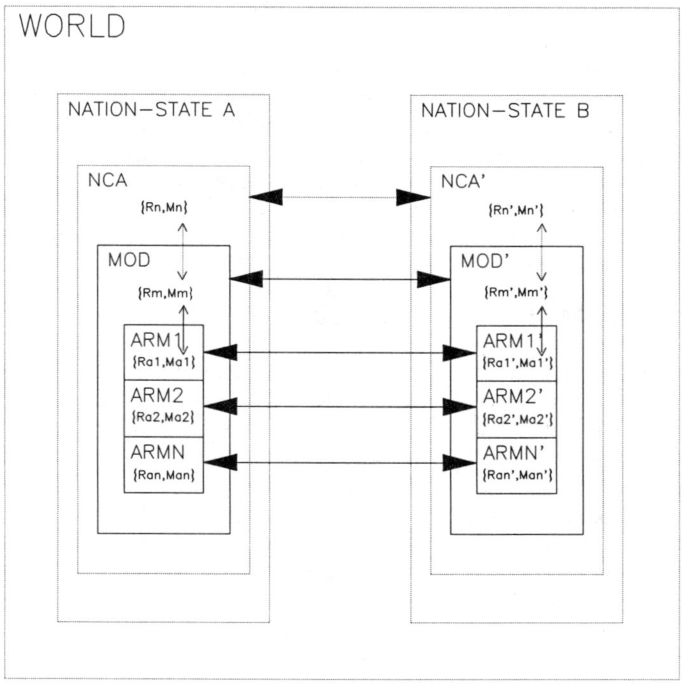

Arrows pointing down transmit goals.
Arrows pointing up transmit capabilities.
Horizontal arrows transmit environmental data.
{Rx,Mx} indicate rules and meta–rules for Unit x.
Dashed lines indicate concepts not included in this model.

The crux of the limitations upon this formalism can be summarized as the problem of reflectivity, and the related question of the production of subjectivity. How this works can be seen by briefly considering the transition function (Z) of a finite state machine. The transition function in a finite-state machine of this type defines a set of relationships among the input to the machine, the state of the machine, and the output of the machine. The specification of the transition function is fixed: It describes the rules, and those are the rules that define the machine. "System transformation" means just this: The rules defining the input-state-output relations change. When its definition changes, it becomes something it might have been instead of what it was.

How could how this happen? Several processes have been proposed, such as trial and error and "meta-knowledge." The second of these is of more interest here. If a machine is endowed with meta-knowledge, we say that it has knowledge not merely of what it is supposed to do but also why it is supposed to do it. The model must, in other words, be equipped first with rules specifying how to act in certain situations and, second, it must know enough about the context to be able to devise new rules when the situation changes. When its input-state-output relationship does not produce the desired result, it has to change that relationship. This is a quality we call "reflectivity." The finite-state machine, by definition, contains no such machinery. Thus, it becomes clear that finite state machine models cannot out-perform the traditional algebraic models on this crucial performance dimension.

5.5 Adaptive Models of Arms Accumulation

The approaches to modeling strategies of arms accumulation reviewed so far--Lambelet's Dreadnought Race model to represent structural models and Etcheson's ARES model to represent production system models--fail to produce satisfying results. The so-called structural or Richardson process models do a fair job of describing synchronic or processual change, i.e., change within highly restrictive structural, temporal, and spatial boundaries. Models like ARES expand the spatial and temporal boundaries, allowing representation of the intra-organizational interactions conceptually necessary for a meaningful understanding of strategy and of system adaptation within the confines of a given structural regime. But neither approach contains representations of structural change, or diachrony. Synchronic change is presumably a crucial issue in the beginnings and endings of particular arms races, as is seen in the case of the Dreadnought Race. Diachronic change, however, seems to be important in establishing the international structures within which arms race behavior may occur in the first place. This section pursues possible research strategies for coming to terms with this issue.

5.5.1 Meta-Rules, Reflectivity, and Second Order Adaptation

The model presented in Chapter Four integrates important features of two recent dissertations on arms accumulation behavior. Squires' model disaggregated arms accumulation processes "horizontally" by examining individual weapons categories.[21] Majeski focused on a "vertical" disaggregation of arms accumulation processes, taking into account different bureaucratic actors in the defense decision-making process.[22] Elements of both are present in ARES, which disaggregates horizontally by military service and vertically by levels of strategic planning. ARES accomplishes this using a formalism that is much simpler than the traditional formulations. The relative simplicity of the production systems formalism thus encourages the modeler to include more information in the model.

With this capability to implement an arbitrarily large amount of detail, there is ample opportunity to devise models incorporating additional explanatory elements beyond those represented in ARES. For example, we may wish to include various insights on organizational behavior provided by rationalist, structuralist and interactionist decision-making theories. Or it might be deemed desirable to include insights offered in other recent studies of arms accumulation behavior. The point to be made is that the simpler production systems approach gives the modeler a far greater degree of discretion for integrating ideas deemed relevant to the explanation of arms accumulation processes than has previously been the case.

Schrodt's thesis examines distributions of power in international systems under the conditions of a Richardson arms race.[23] His central question is to determine the conditions under which categories of arms distributions are maintained in the face of changing levels of arms. Focusing on the preservation of various balance-of-power regimes, one question left unanswered in Schrodt's work concerns the timing and force levels in cases where an international system does not preserve the status quo regime. Schrodt remarks that "it would be interesting to see whether methods simpler than the solution of differential equations can be used to study this problem."[24] The ARES model presented in the previous chapter suggests an affirmative answer to this query. We will proceed in the remainder of this chapter to illustrate the capability of the production systems formalism to incorporate arbitrary degrees of complexity by focusing on this particular problem. We will begin this task by mentioning a concrete historical case of diachronic change in the international system.

Internal change in nation-states and external change in the international system are closely linked. Prussia gained a new political stature on the continent following the wars of unification, allowing Bismark to be an "honest broker" in the African land-grab at the Berlin Conference of 1884. The Sun King's extravagant failure in the War of Spanish Succession exhausted French resources and endowed Britain with material and political capital for maintaining hegemony through the eighteenth century. These kinds of internal changes in the nation-state have far-reaching implications for the structure of the international system and for the behavior of states within the system. As international resource distributions and prevailing intersocietal conventions change through evolution and/or war, a new matrix of international possibilities arises. After 1945, for example, Great Britain no longer controlled the resources adequate to maintain greatness. What changed? As the balance of capabilities in the international system changed, thus did the rules of the game change. This is where arms race parameters come from.

Intelligence, self-knowledge, and reflectivity are required to manage state policy in the face of diachronic system change. It required a measure of intelligence for Winston Churchill's successors to "preside over the dissolution of the empire," beginning with their move to bow out of the Greek civil war in 1946. It required reflectivity to realize that the new means were no longer adequate to or appropriate for the old goals. The centuries-old "two power plus 10%" naval standard was quite simply not practical in the face of the new-found American and Soviet power. It required knowledge of the overall structure of Britain's foreign policy goals--self-knowledge--to redefine the scope of the nation's ends-means matrix and successfully effect the transition from hegemon to second-tier power.

This is not to say that such changes happen all at once. Learning new rules takes time for organizations no less than for people. The fiasco of the 1956 Suez crisis is a case in point. The British deployed an old and trusty plan in a new environment, but it didn't work like it used to. The old plan did not fit the new world, so new plans would have to be generated to fit the new world if the state was to survive.

5.5.2 Concepts for Models of Adaptive Arms Accumulation

The case mentioned above by Schrodt is a situation of diachronic change in the international system, a problem notoriously difficult to study employing the standard formalisms of the Richardson Tradition. One can begin to approach the problem of modeling diachronic change in arms accumulation processes by applying a set of concepts called "meta-rules," or rules about rules. Recent research in the field of artificial intelligence has developed this concept for application to automated planning processes. Actually, the idea of automated planning is not exactly brand new; Perseus, son of the Greek god Zeus, had a golden automatic planning machine built in the form of a wise owl. The contemporary efforts have been less successful, if more applicable to the case at hand.

The roots of scientific studies of automated planning stretch back only to the late 1950s and early 1960s.[25] Newell, J.C. Shaw, and Simon put forward the production system formalism and implemented it in their early planner, GPS.[26] This early planner was, however, linear and lacking in the ability to deal with the complexity of real world planning tasks. The idea of meta-level system control was first implemented by Randall Davis in the TIERESIAS program, discussed at length in Chapter Three.[27] The first successful implementation of a truly automated planning system was Earl Sacerdoti, who specified a system for planning electro-mechanical assembly tasks.[28] Michael Zisman's automated office procedures formalism is typical of the class of programs inspired by Sacerdoti's work.[29] They are limited by their domain rigidity and task specificity.

Work is continuing in the AI community on a variety of more sophisticated planning formalisms. Some of the active lines of research are suggested in the work of Barbara Hayes-Roth et al. on incremental opportunistic planning,[30] Daniel Corkill on hierarchical planning,[31] John McDermott on learning to use analogies,[32] and Sacerdoti on problem-solving tactics.[33] A particularly notable example of such research, one with direct relevance to the study of politics, is provided by the work of Carbonell, who models counter-planning under adversity.[34]

The bulk of these research efforts are in their early stages of development. For the most part, the resulting formalisms are not yet sufficiently well developed to support just any application. Models of particular planning tasks must still be constructed from scratch, and thus the development of large-scale planning systems is a highly labor-intensive process. Nonetheless, artificial intelligence research continues to uncover new possibilities for the representation and manipulation of the empirical and theoretical subject matter. International relations should remain alert to these developments and move to exploit existing techniques.

It is possible to outline an approach to the modeling of diachrony in arms accumulation processes relying on these techniques. The substantive material remains underdeveloped at this time, and is presented only as an illustration of the approach.

Let us consider the evolution of British arms accumulation policy in this century. Implementation of their "two power plus 10%" standard for naval construction required ongoing monitoring of the environment. In order to generate appropriate rules for synchronic change in the supply rules of the military services, the British MOD must scan the environment to find the two largest naval powers, upon which the "two power" standard is to be keyed. Let's call this scanning process the "MOD planner." It can be represented as a simple algorithm or set of rules:

(1) Create a master list of all naval powers
(2) Delete smaller of first two nations on the list
(3) Repeat (2) until only one navy remains
(4) Add that navy to the key list, delete it from master
(5) Repeat 1-4 to find second largest navy
(6) Calculate naval supply functions on key list.

This simple procedure creates a "key list" of the two largest navies, from which the two power standard supply functions may be calculated. The MOD planner can be implemented easily as a first-order production system.

Using the MOD planner, the transition into the beginning of the Anglo-German naval competition is fairly straightforward. Prior to the 1901-1905 transition period, the two power standard had been calculated upon the two largest foreign fleets, the French and the Russian. As the German building plans began to bear fruit, however, it became clear that soon the German fleet would overtake the Russian fleet in size. The MOD planner would detect this and make the requisite changes in supply functions.

Negotiations at the 1922 Washington Naval Conference led to a formula that defined a five-to-one-to-three ratio of U.S., British, and Japanese capital warships. While this would appear on the face of it to mark the end of the British two power standard, this is not so. Competition in warships simply shifted to other classes, including light cruisers, destroyers, and submarines.

The end of the World War II, however, is another matter. With the emergence of the Soviet Union and the United States as the dominant powers of the international system, the two-power standard was no longer applicable. It therefore seems that the structure of the rule-generating process in Britain had to change around that time. This is where the meta-rule concept becomes useful. One can illustrate how such a set of changes might be initiated by defining a high-level meta-rule as part of the Britain's NCA. Let's call it Meta-Rule 1, or simply "M1":

```
[M1: Define International System Structure

      (Use Master List of All Military Powers, ordered by size)
(COND ((Largest Power is greater than second+third+fourth Powers)
            (International System is Uni-polar))
      ((Second Power is greater than third+fourth+fifth Powers)
            (International System is Bi-polar))
      (ELSE (International System is Multi-polar))
**Polarity Sub-rules:
(COND ((International System is Uni-polar)
            (AND (Britain is the Pole))
                  (Apply Balancer Rules)
            (ELSE (Apply Alliance Rules)))
      ((International System is Bi-polar)
            (AND (Britain is a Pole))
                  (Apply Growth Rules)
            (ELSE (Apply Alliance Rules)))
      ((International System is Multi-polar)
                  (Apply Growth Rules))]
```

And so on. Thus, M1 would control a larger set of subordinate rules, i.e., the MOD planning rules that in turn generate such rules as the naval supply functions. One such set of planning rules is here called the "balancer rules," which refers to that set of rules held to apply when Britain is hegemonic in the system. The balancer rules would include such things as instructing the MOD planner to define and implement the two-power standard by relying on sea power.

5.5.3 Analysis of Adaptive Approaches

Although much detail remains to be worked out in order to implement the adaptive model sketched above, at first glance the approach does seem to offer some possibility for advancing toward a conceptually adequate model of armaments accumulation processes. All that remains, seemingly, are the details of implementation. Just specify an adequate set of meta-rules, implement them in code, test them against some empirical data, and by retroduction refine them into an empirically faithful simulation of the world. Bubbling with optimism, the modeler presents the achievement to a packed, hushed auditorium, when suddenly a voice from the back shatters the expectant quietude. Echoing the challenge by modeler to the structural approach the voice demands, "Just where did all these (meta-) rules come from?" Gasping with resentment, modeler blurts, "Why, from my observation!" *Ex post facto*. The intentionality embodied in the model is that of the modeler, not that of the model. The modeler is confronted with the spectre of infinite regress. Specifying more rules to govern change in the meta-rules does not help. The rules, if they are to reflect the way it really happens, must change themselves as a function of what they are and potentially are. How do we conceptualize this? That is the $64,000 social science bonus question.

5.6 Summary on Strategies of Arms Accumulation Research

Several significant points stand out from the foregoing discussion.

§ *Richardson sought emancipatory solutions to international security problems.*

§ *Such a goal entails attention to meta-level structures.*

§ *The structural and first-order production system models put forward by Richardson and his students have contained neither representations nor conceptualizations of these necessary meta-level structures.*

§ *A method for attaining this representation is proposed, namely, second-order production system models of strategic evolution.*

Assuming a view of international conflict as an interactive system-level epiphenomenon, the Richardson Tradition of armaments accumulation analysis has sought to define and elaborate emancipatory solutions to insecurity dilemmas. We assert that a technical capability to engage in discourse about the meta-level structural dynamics of conflict systems is a prerequisite to realization of the goals of the Richardson Tradition. Structural and first-order production system models of arms accumulation have been examined and found wanting in meta-level structures.

A method has been proposed for implementing such a technical capability. Armaments accumulation processes are conceptualized as planning behaviors, facilitating theorizing about linkages between strategic interaction and meta-level or diachronic change in national and international systems. Conceptualized as a network of planning behaviors, strategic interaction in armaments accumulation processes can be analyzed for meta-level properties associated with system transformation. This promising approach warrents further investigation. However, we conclude that the technical ability to engage in meta-level discourse is a necessary, but not a sufficient, condition for realization of conceptually adequate models of arms accumulation behavior.

Notes

1. G.A. Miller, E. Galanter, and K.H. Pribram, Plans and the Structure of Behavior (New York: Holt, Rinehart and Winston, 1960), p. 16.

2. B.H. Liddell-Hart, Strategy (New York: Signet, 1967), p. 322.

3. Karl Von Clausewitz, On War, trans. Anatol Rapoport (Princeton, NJ: Princeton University Press, 1976), p. 241.

4. Abba Eban, The New Diplomacy: International Affairs in the Modern Age (New York: Random House, 1983), p. 388.

5. Liddell-Hart, Strategy, op. cit., p. 320.

6. Howard Tamashiro, "Algorithms, Heuristics, and Artificial Intelligence Modelling of Strategic Statecraft," in Foreign Policy Decision Making: Perception, Cognition, and Artificial Intelligence, ed. Donald A. Sylvan and Steve Chan (New York: Praeger, 1984), p. 200; cf. Edward Luttwak, Strategy and Politics (New Brunswick, NJ: Transaction, 1980).

7. Martin Van Creveld, Command in War (Cambridge, MA: Harvard University Press, 1985), p. 58.

8. E.L. Woodward, Great Britain and the German Navy (Oxford, England: Claredon Press, 1935), p. 21.

9. Felix Gilbert, The End of the European Era, 1890 to the Present (New York: Norton, 1970), p. 72.

10. Woodward, Great Britain and the German Navy, op. cit.

11. Margaret Tutle Sprout, "Mahan: Evangelist of Sea Power," in Makers of Modern Strategy: Military Thought from Machiavelli to Hitler, ed. E.M. Earle (Princeton, NJ: Princeton University Press, 1971), p. 442.

12. Quoted in Nazli Choucri and Robert C. North, Nations in Conflict: National Growth and International Violence (San Francisco: W.H. Freeman and Co., 1975), p. 105.

13. Jean-Christian Lambelet, beginning with his historical analysis, "The Anglo-German Dreadnought Race," Peace Science Society (International) Papers 22 (1974); moving into his data-gathering phase with "A Numerical Model of the Anglo-German Dreadnought Race," Peace Science Society (International) Papers 24 (1975), pp. 29-48; and finally the summary analysis, "A Complementary Analysis of the Anglo-German Dreadnought Race, 1905-1916," Peace Science Society (International) Papers 26 (1976), pp. 49-66.

14. Michael D. Intriligator and Dagobert Brito, "Heuristic Decision Rules, the Dynamics of the Arms Race, and War Initiation," paper presented to the conference on Dynamic Models of Conflict, Boulder, CO, October 31 to November 3, 1984, p. 6; emphasis in original.

15. Von Clausewitz, On War, op. cit., p. 402.

16. Choucri and North, Nations in Conflict, op. cit., pp. 115-117.

17. Lambelet, "The Anglo-German Dreadnought Race," op. cit., p. 44.

18. Craig Etcheson, "Strategy in the Arms Race: A Qualitative Computer Simulation," paper presented to the annual meeting of the International Studies Association, Washington, DC, March 5-9, 1985; and Chapter Four of the present work.

19. Avron Barr and Edward A. Feigenbaum, The Handbook of Artificial Intelligence, 3 vols. (Stanford, CA: HeurisTech Press, 1982).

20. Paul A. Anderson and Stuart J. Thorson, "Artificial Intelligence Based Simulation of Foreign Policy Decision-Making," Behavioral Science 27 (April 1982), pp. 176-193; Stuart J. Thorson and Donald A. Sylvan, "Counterfactuals and the Cuban Missile Crisis," International Studies Quarterly 26 (December 1982), pp. 539-571; and Stuart J. Thorson, "Intentional Inferencing in Foreign Policy: An AI Approach," in Foreign Policy Decision Making: Perception, Cognition, and Artificial Intelligence, ed. Donald A. Sylvan and Steve Chan (New York: Praeger, 1984), pp. 280-309.

21. Michael L. Squires, "Modeling the U.S.-U.S.S.R. Strategic Arms Race" (Ph.D. dissertation, Indiana University, 1982).

22. Stephen J. Majeski, "Mathematical Models of the U.S. Military Expenditure Decision-Making Process" (Ph.D. dissertation, Indiana University, 1981).

23. Philip A. Schrodt, "The Effects of a Richardson Arms Race on the Distribution of Power in an International System" (Ph.D. dissertation, Indiana University, 1976).

24. Ibid., p. 279.

25. Alan Newell, J.C. Shaw, and Herbert A. Simon, "Chess-Playing Programs and the Problem of Complexity," IBM Journal of Research and Development 2 (October 1958), pp. 320-335; Miller, Galanter, and Pribram, Plans and the Structure of Behavior, op. cit.

26. Alan Newell and Herbert A. Simon, Human Problem Solving (Englewood Cliffs, NJ: Prentice-Hall, 1972).

27. Randall Davis, "Applications of Meta-Level Knowledge to the Construction, Maintenance and Use of Large Knowledge Bases" (Ph.D. dissertation, Stanford University, 1976).

28. Earl Sacerdoti, A Structure for Plans and Behavior (New York: Elsevier, 1977).

29. Michael D. Zisman, "Representation, Specification, and Automation of Office Procedures" (Ph.D. dissertation, University of Pennsylvania, 1977).

30. Barbara Hayes-Roth, Frederick Hayes-Roth, Stan Rosenschein, and Stephanie Cammarata, "Modeling Planning as an Incremental, Opportunistic Process," Proceedings of the sixth International Joint Conference on Artificial Intelligence, Tokyo, August 20-23, 1979, Vol. 1, pp. 375-383.

31. Daniel D. Corkill, "Hierarchical Planning in a Distributed Environment," Proceedings of the sixth International Joint Conference on Artificial Intelligence, Tokyo, August 20-23, 1979, Vol. 1, pp. 168-175.

32. John McDermott, "Learning to Use Analogies," Proceedings of the sixth International Joint Conference on Artificial Intelligence, Tokyo, August 20-23, 1979, Vol. 1, pp. 568-576.

33. Earl Sacerdoti, "Problem Solving Tactics," Proceedings of the sixth International Joint Conference on Artificial Intelligence, Tokyo, August 20-23, 1979, Vol. 2, pp. 1077-1085.

34. Jamie Carbonell, "The Counterplanning Process: A Model of Decision-Making in Adverse Situations," Pittsburgh: Carnegie-Mellon University, Department of Computer Science, February 1979.

6 SUMMARY AND CONCLUSIONS ON ARMS ACCUMULATION

For a long time, preparation for and conduct of war has served as a central institution of human societies. At the present stage of historical development, however, it seems likely that another conflict on the scale and at the scope of this century's global actions would not serve the interests of any of the advanced industrial nation-states, the primary belligerents in any such general conflict. In fact, it is not at all inconceivable that a large-scale nuclear war would bring to an end the state-centric system as it has coexisted with Western civilization since the state system emerged from the ruins of the Holy Roman Empire. If the prognosticators of the "nuclear winter" theory are even close to correct, this is made certain.

Thus, at the present stage of international development a global fight might not even serve the interests of the international system as a whole in the sense that war has served to maintain the state-centric system. Therefore, if one values the maintenance of an international system that looks at least remotely similar to the present one, it seems that the avoidance of overt general war is a worthy goal. Stable, intelligently managed, and predictable change in the strategic arsenals of the major powers, most agree, will help in this avoidance effort. How to do this is the question.

6.1 Understanding Strategic Interaction Is Important

This research effort has been concerned with attempts to understand the evolution of military bureaucracies and with the interactions of the bureaucracies with each other and with the observer. What has not been done, however, is pause to ask why this is an important question. This section addresses my concern with understanding what controls change in military organizations. I believe that such an understanding is of essential importance. If one has no clear analytical idea of how large modern military bureaucracies change, then proposals to change the institutions can hardly be founded upon solid ground. And there are countless such proposals. Some of them may have merit. Most probably don't. Many are contradictory. Which is which?

For example, in the decade of the 1980s, the "nuclear freeze" proposal--or rather proposals, for it is of many shades--has been much bandied about. In the lurch, Alton Fry suggests the "build down" proposal. Ronald Reagan offers the "strategic defense initiative." Mikhail Gorbachev offers the complement, a "star wars ban." These are only a few of the recent proposals, those receiving the most media attention. Many more lurk in the halls of government and academia, waiting to be sprung upon eager audiences everywhere. Oddly enough, nearly all of these widely disparate proposals claim to have precisely the same goal: to reduce dependence upon nuclear weapons while enhancing the security of the nation and, presumably, of the world.

One significant difficultly is that we have no systematic way to assess the merit of these proposals. There is no scientific way to predict the consequences of implementing any one of these widely disparate plans for altering the shape of military bureaucracies. Perhaps one (or more) of these proposals is, or is close to, the answer sought. In the heat of partisan politics, politicians will be quick to claim that recent history confirms their particular view, but will the historians of one hundred years hence see it the same way? One of these proposals could prevent World War III. Or cause it. But which is which? We musn't have the patience to wait for future scribblers to find out. The development of approaches to address these kinds of questions constitutes an important problem and challenge. This is the problem the present research attempts to address, if in a tentative and exploratory manner. But this certainly is not a new problem. It was precisely this problem that propelled Lewis Fry Richardson on his lifelong mission.

6.2 Richardson Sought Emancipatory Solutions

Perhaps the cardinal virtue of Lewis Fry Richardson's labor is one little discussed by his students. It derives from his definition of the "warpeace" particle: the "outward attitude of threatening or cooperation." It is important to note the cooperation part. Though conflict seemed to be the force that activated Richardson, the possibility of peace was the force that motivated him. The possibility of peace, indeed, is inherent in his design. Most representations of the international state space in the Richardson Tradition are restricted to only the northeast quadrant of the space, where the value of the warpeace particle is positive for both nations. Positive values of the warpeace particle signify conflict. This quadrant has been the primary focus of analytic activity among Richardson's students. Negative values of the warpeace particle, located in the third quadrant of the international space, by contrast, denote concord. Richardson tried to operationalize this concept as international trade, but his results are problematic and his students have tended to avoid the entire issue.

We always speak of the "conflict spiral," but one never hears of a "peace spiral."[1] Is that because of what we observe? Or perhaps because of what we expect to observe? By and large, Richardson's students have not honored the full scope of his vision. The Richardson Tradition of armaments accumulation analysis should rededicate itself to Richardson's vision. This means that the tradition must focus more attention upon attempts to find emancipatory strategies that would lead out of conflictual quadrants of the international space.

The way to do this does not involve abandoning pell-mell the metaphysical basis of positivism, but it does require a re-examination of how Richardson and his students have applied the tenets of positivism in their efforts to fashion a science of peace. Happily, this re-examination seems to be well under way. It involves examining our methodological bases and looking for ways to improve our ability to study these elusive questions. Using the most modern mathematical methods available does not seem like too much to recommend. After all, some of the most innovative and flexible computer simulation techniques yet created are being used to plan the fighting of World

War III. This same technology might be usefully applied in service of the ends of peace research. As this study has shown by example, however, we must beware the false notion that the technologies we use to study the problem can in and of themselves solve it. Solutions will require our imagination and our dedication.

6.3 Strategies for the Fifth Phase

It is the task of the social scientist and the peace researcher to understand the past and the present well enough to know a little bit about what is going to happen in the future, enough so that one can help prepare the present for what is to come. Driven by that which lies behind, peace research is defined by the tasks that lie before it. These tasks can be divided into those for the short term and those for the long term. In the short term, the task of peace research is to develop ideas and strategies toward mitigating unnecessary violence in human social and political affairs. In the long term, the task is more formidable but not less urgent: to develop basic scientific knowledge that uncovers the ontological structures giving rise to pervasive and highly organized human institutions of social violence, with a view to imagining possible futures wherein the structural bases of social violence are transformed so as to lessen the occurrence of unnecessary violence. In this sense the student of Lewis Fry Richardson is following along in the footsteps of the founder of scientific peace research.

Both of these tasks confronting peace research, the short term and the long, necessarily imply a pedagogical function. This function must be distinguished from a propaganda function, though this is a difficult task. This is doubly difficult in the Richardson Tradition because the Richardson model is so vulnerable to misuse in the classroom, to being presented in such a way as to reinforce invariant, ahistorical views of international relations. "Teaching peace" is logically not only necessary but elemental to the success of peace research. It is an idealistic enterprise to be sure. But as a wise teacher once said, the only rational response to impossible odds is to try harder.

As Richardson, Rapoport, Alker, Ashley, and others have argued, we must seek emancipatory perspectives on the phenomenon of armaments accumulation. Positive science traditionally gives equal emphasis to its two driving goals, control and understanding. We must remember that the former depends entirely on the latter. In the Richardson Tradition scientific knowledge has not been easy to pry from Nature. The gods continue to guard important bits of wisdom jealously.

Such a state of affairs doubtless will propell a new generation of Richardson's students after the "secrets of the Old One," creating a fifth phase in the epistemological history of the Richardson Tradition. After the mutuality, autonomy, mixed, and teleological models of the first four phases, perhaps the models of the fifth phase can be characterized as cognitive models. As Richard Hermann has recently argued, one of the principal challenges of the "cognitive revolution" in the social sciences is "to restructure theoretical conceptions in a way that we encompass the complexity introduced by treating humans as subjects rather than as scientific objects."[2] This challenge parallels the challenges facing the Richardson Tradition. If this is correct, the fifth phase of the Richardson Tradition will seek to incorporate notions of intentionality, pulling on the scholarship of the "cognitive revolution," the hermeneutic tradition, and developments in anthropology, sociology, and linguistics; it will reflect the developing philosophical consequences of quantum mechanics upon our understanding of what we can and do know about what we know; and it will reflect the holistic, interconnected, probabalistic, and contingent, observer-created reality perspectives sweeping all of the sciences. And it will increasingly reflect the influences of the sciences of the artificial.

6.4 The Limits of Innovation in Arms Modeling

Artificial intelligence-inspired works in political science and international relations will enjoy considerable vogue in the coming decade. This stands in marked contrast to the situation only a few years ago. When the outlines of the present project were taking shape in 1977 and 1978, the political science literature contained very few references to artificial intelligence. With the exception of Alker and a couple of his students,[3] such references as did exist were limited to information scientists with a side interest in politics[4] and the odd piece in military science.[5] Over the decade-long course of the present study, however, this situation has changed in significant and dramatic ways.

The first half of the 1980s witnesses the emergence of a group of young political scientists interested in the development of AI applications. The list of individuals involved in this movement includes Paul Anderson, Bennett, Steve Chan, C. Christensen, Ensign, Etcheson, Majeski, Mefford, Miller, Schrodt, the Sylvan cousins, Tamashiro, Akihiko Tanaka, and Thorson, to mention just some of those who have published on the topic. Is this the makings of a fad?

This is not an unreasonable question; AI itself was considered a fad within the computer science community through the 1960s and 1970s. That it is now a multi-billion dollar industry to some extent vindicates early AI proponents from the "fad" label, but as Schrodt and many others have pointed out, AI does have a long tradition of failing to deliver on its promises.[6] Much of the most serious criticism of AI, moreover, has been and continues to be directed not at its commercial potential but rather at some of AI's more glorious long-term aspirations.[7] But what about AI applications in the field of international relations?

Take a look at the list of international relations scholars who are interested in AI applications. What is particularly significant in light of the present study is the close identity of interest among this group. Besides the fact that many of those listed are students of Alker, one thing binding the group is a wide interest in international conflict. Several of them have done extensive work on arms race modeling, including the two Zinnes students (Schrodt and Majeski) whose dissertations were in the Richardson Tradition. Why do so many scholars interested in international conflict seem to be "discovering" AI?

6.4.1 The Potential of AI Techniques

The names of many students of arms races appear on a list of those now pursuing AI-oriented work. The present research has shown that the methodological challenges of the Richardson Tradition are sufficient to drive its practitioners in search of new methods. But why have so many gravitated to AI? It seems that there may be a set of reasons that lie at the interface of theory and method.

Any serious recitation of problems plaguing work in the Richardson Tradition would include several items related to the dominant classical method: (1) the problem of diachrony, or the structural invariance of econometric models; (2) the problem of knowledge representation, or the limited data-handling capacities of the quantitative metric; (3) the problem of estimation, i.e., the elusiveness of confident empirical findings; and (4) the problem of non-Von Neumann architectures, i.e., the inherent linearity of the General Linear Model. All of these problems appear to be mitigated or even eliminated in AI approaches. This is no small breakthrough in and of itself and may be sufficient to account for the gathering interest in AI.

Schrodt's review of the potential advantages of an AI approach points out some theoretical reasons for this growing interest. He notes that AI is characterized as data intensive, non-numerical, computer intensive, flexible with respect to rationality, flexible with respect to stochastic structures and estimation, and formal and intersubjective.[8] All of these characteristics of AI, Schrodt contends, are congenial to the subject matter of international relations in general and armaments accumulation studies in particular. The thrust of this study has supported Schrodt's contention. How does the promise measure up so far?

6.4.2 Limits to the Scope of the AI Contribution

There now are enough examples of AI applications in international relations to warrant a tentative catalogue. A serious evaluation of this emerging literature will be appropriate in the next report of this on-going research program. Nonetheless, it is possible to identify several emerging areas of concentration in AI-inspired international relations applications. The extant work seems to cluster in three areas: methodology, simulation, and pattern matching. The first of these categories has produced several pieces that are largely methodological essays. These focus on various questions concerned with the application of AI methods to international relations.[9]

The dominant concentration of AI work in international relations, however, seems to lie in the construction of simulation machines. This is leading to the proliferation of a new crop of acronyms, such as Carbonell's POLITICS,[10] Thorson's JFK/CBA and PIE,[11] Tanaka's CHINA_WATCHER,[12] and Etcheson's CARREEN[13] and ARES.[14] Other AI-inspired simulation work includes Alker's early contributions of a UN peace-making simulation with a "teaspoon of artificial intelligence"[15] and the War of the Pacific Simulation, which included a rudimentary form of precedent-based decision making.[16]

A third area of AI concentration in international relations is that concerned with pattern matching. More precisely, work in this area tends to follow along the lines of research on natural language pattern matching.[17] It is possible to speculate that when the work inspired by pattern matching research is fully married to the work on simulation systems, the basis will exist for some extremely interesting AI applications. What other imaginative applications may come down the pike in the future remain to be seen.

These three areas of concentration--methods, simulation, and pattern matching--do not exhaust the possible applications of AI in international relations. In each of these emerging areas of concentration, research remains in tentative and exploratory phases. There is a danger of expecting too much too soon from this exciting technology, but this must be balanced with a careful examination of just what the possibilities are. This will inevitably entail marching resolutely down many blind alleys, bravely off the edges of many precipices, and, if necessary at times, flat into some brick walls. The present study is perhaps a case in point. Eyes should be open to the fact that this technology cannot solve the conceptual and theoretical problems, which are prior and independent. Just as the growing use of quantitative methods in social science led to abuses, AI techniques are open to abuse. There will in all likelihood soon emerge an AI equilvalent to "mindless number crunching." In fact, this danger may even be greater using AI approaches due to the generally unstructured or loosely structured nature of many AI techniques. There will probably be much ill-conceived foolishness before the wheat is separated from the chaff. Neither quantitative nor qualitative methods by themselves ensure rigor.

Another problem of expecting too much too soon from AI applications in international relations is that, for much progress to be made in the larger picture, the field would really have to undergo a massive retooling, a retooling too massive to expect any time soon. People in political science have spent the last twenty-five years getting used to teaching statistics and stressing positive and structuralist perspectives in training members of the discipline. Now we are going to insist on training in heuristic methods instead of or in addition to algorithmic methods? Perhaps the requirements of retooling would be somewhat mitigated by the friendly user-oriented qualitities of some higher level AI research tools. In part this also may be a generational process. In Alker's generation, he was more or less a lone pioneer, no doubt often thought of as more than a bit off the beaten path. In the emerging generation, there are a score or more researchers with AI orientations. How many will the current generation train? If the resolution of epistemological crises of degenerative problem shifts is a generational process, then this question will have a direct bearing on scientific progress in the Richardson Tradition of armaments accumulation analysis.

6.5 Future Research

Drew McDermott must surely have been anticipating the current phases of the research project reported in this volume when he wrote

> Here, however, he begins to lose his grip. Implementing Version I, whose shortcomings are all too obvious, was exhausting; it made him feel grubby for nothing. (Not at all like the TECO macros he took time out for along the way!) He feels as though he's paid his dues; now he can join the theoreticians. What's more, he should. Implementation details will make his thesis dull. The people want epistemology. Simultaneously, he enjoys the contradictory feeling that the implementation of Version II would be easy. . . .[18]

The ARES simulation system has explored the applicability of elementary AI-type data structures and control structures to problems of interaction in armaments accumulation. It does little more, having neither empirical content nor historical meaning. With the initial effort complete, the project must move to a more substantive case study, as well as more sophisticated control structures, in order to sustain progress. This would involve a case with specific and known empirical content, preferably a case that has had the benefit of extensive prior investigation. The project must move beyond "Version I" if it is to prove of lasting value.

In fact, work on Version II has already begun. This model is tentatively called "SPIRAL" (strategic planning in reciprocal armaments accumulation). The substantive case is one much studied and well documented, the pre-World War I Anglo-German Dreadnought Race. A central design feature of SPIRAL will be the use of a control structure based upon meta-rules. This focus will allow the implementation of more realistic and elaborate strategic planning capabilities, such as those outlined in the previous chapter. SPIRAL should have empirical content, as well, which implies a need for some measures of validity. The construction and validation of this second generation system will not be easy.

Experience gained in the construction of the ARES and SPIRAL models is aimed at building a base of support for an eventual assault on the primary objective of this project. That would be the construction of a machine that behaves like the real things--the ministries of defense of the contemporary world. As the content of the present research effort has graphically demonstrated, it is no small task to model such things. In fact, there remain fundamental conceptual barriers to the accomplishment of this aim. This does not, however, diminish the urgency of the goal, that is, improving our understanding of strategy and the structure of behavior in armaments accumulation processes.

Classical analysis in the Richardson Tradition has meant attempting to define the path and velocity of the warpeace particle in the warpeace space. Once the behavior of this fundamental particle was known with certainty, so the Newtonians like Richardson believed, all causes and effects of the system, future and past, could then be extrapolated. But in the problem domain of particle physics, Werner Heisenberg destroyed this idea with his Uncertainty Principle. And today, the subsequent quantum mechanical revolution that has swept away classical physics in this century rejects the view that such an "elementalist" or "atomistic" or "reductionist" approach can ever lead to understanding. We can say that we see a move toward this kind of more holistic view in the Richardson Tradition as well. Through a series of epistemological crises, the Richardson Tradition has come around to the notion that the warpeace particle in the international state space is impelled not simply by its momentum, but that this particle is somehow active, intelligent, and capable of altering its trajectory in the state space--indeed, even the shape of the state space itself--at will. This is what we are challenged to capture in the next generation of our theories and models of armaments accumulation.

6.6 Final Summary and Overall Conclusions

We have learned several things.

§ *The military interaction of nation-states, though much studied, remains highly puzzling and eminently in need of lucid explanation that supports understanding.*

§ *The Richardson Tradition of armaments accumulation studies has moved steadily toward explanations incorporating elements of strategy and bureaucracy.*

§ *We have yet to identify the mathematical structures describing the motion of Richardson's warpeace particle in the international plane.*

§ *The first-order production systems formalism implemented in ARES cannot yield a conceptually adequate model of armaments accumulation, though more sophisticated implementations of this technology offer a largely unexplored and potentially fruitful avenue of exploration.*

§ *Technology alone will prove insufficient; we must draw most of all on our creativity and imagination to conceptualize emancipatory perspectives on the social-structural problems of peace reseach.*

Emancipatory solutions to the dilemmas facing humankind lie within our selves; whether we shall ever grasp this potential depends, at least in part, on the extent to which we are able to improve our understanding of strategy and the structure of behavior in arms races.

Notes

1. There are a few recent and laudable exceptions, for example, Dina Zinnes, "Modeling International Cooperation," paper presented to the annual meeting of the International Studies Association, Washington, DC, April 14-18, 1987.

2. Richard Hermann, "The Empirical Challenge of the Cognitive Revolution: A Strategy for Drawing Inferences About Perceptions," International Studies Quarterly 32 (1988), pp. 175-203.

3. Hayward R. Alker, Jr., and C. Christensen, "From Causal Modeling to Artificial Intelligence: The Evolution of a UN Peace-Making Simulation," in Experimentation and Simulation in Political Science, ed. J.A. LaPonce and Paul Smoker (Toronto: Toronto University Press, 1972); and also James P. Bennett and Hayward R. Alker, Jr., "When Collective Security Policies Bred Collective Insecurity: The War of the Pacific in a World Politics Simulation," in Problems of World Modeling, ed. Karl W. Deutsch, Bruno Fritsch, Helio Jaguaribe, and Andrei Markovits (Cambridge, MA: Ballinger Publishing Co., 1977), pp. 215-302.

4. Jamie G. Carbonell, "POLITICS: Automated Ideological Reasoning," *Cognitive Science* 2 (1978), pp. 27-51; also Carbonell's "The Counterplanning Process: A Model of Decision-Making in Adverse Situations," op. cit.; and his "POLITICS," in *Inside Computer Understanding*, ed. R. Schank and C. Riesbeck (Hillsdale, NJ: Lawrence Erlbaum, 1981), pp. 259-307.

5. Marshall Pease III, "ACS.1: An Experimental Automated Command Support System," *IEEE Transactions SMC* 8 (October 1978), pp. 726-735.

6. Drew McDermott, "Artificial Intelligence Meets Natural Stupidity," in *Mind Design: Philosophy, Psychology, Artificial Intelligence*, ed. John Haugeland (Cambridge, MA: MIT Press, 1981), pp. 143-160; and also Philip A. Schrodt, "Artificial Intelligence and the State of Mathematical Modeling in International Relations," paper presented to the Conference on Dynamic Models of Conflict, Boulder, CO, October 31 to November 3, 1984.

7. Hubert L. Dreyfus, "From Micro-Worlds to Knowledge Representation: AI at an Impasse," in *Mind Design: Philosophy, Psychology, Artificial Intelligence*, ed. John Haugeland (Cambridge, MA: MIT Press, 1981), pp. 161-204; John R. Searle, "Minds, Brains, and Programs," also in Haugeland's *Mind Design*, pp. 282-306.

8. Schrodt, "Artificial Intelligence and the State of Mathematical Modeling in International Relations," op. cit.

9. R. Miller and Stuart Thorson, "Control Concepts in Theories of Governmental Decision-Making: A Structural Approach," in *Mathematical Systems in International Relations Research*, ed. John V. Gillespie and Dina A. Zinnes (New York: Praeger, 1977), pp. 60-90; Craig Etcheson and Walt Scacchi, "MONADYSIM: Applying AI Techniques to the Study of War," *Proceedings* of the IASTED International Symposium: Applied Simulation and Modeling (Anaheim, CA: ACTA Press, 1982), pp. 35-38; Schrodt, "Artificial Intelligence and the State of Mathematical Modeling in International Relations," op. cit.; Howard Tamashiro, "Algorithms, Heuristics, and Artificial Intelligence Modelling of Strategic Statecraft," in *Foreign Policy Decision Making: Perception, Cognition, and Artificial Intelligence*, ed. Donald A. Sylvan and Steve Chan (New York: Praeger, 1984), pp. 197-226.

10. Carbonell, "The Counterplanning Process," op. cit., and also his "POLITICS," op. cit.

11. Anderson and Thorson, "Artificial Intelligence Based Simulation of Foreign Policy Decision-Making," op. cit.; Thorson and Sylvan, "Counterfactuals and the Cuban Missile Crisis," op. cit.; Thorson, "Intentional Inferencing in Foreign Policy: An AI Approach," op. cit.

12. Akihiko Tanaka, "China, China Watching, and CHINA_WATCHER," in *Foreign Policy Decision Making: Perception, Cognition, and Artificial Intelligence*, ed. Donald A. Sylvan and Steve Chan (New York: Praeger, 1984), pp. 310-344.

13. Craig Etcheson, "Expert Systems for Country Risk Analysis: Systematizing Qualitative Data in Political Risk Analysis," paper presented to the annual meeting of the International Studies Association, Atlanta, March 27 to April 1, 1984.

14. Etcheson, "Strategy in the Arms Race," op. cit.

15. Alker and Christensen, "From Causal Modeling to Artificial Intelligence," op. cit.

16. Bennett and Alker, "When Collective Security Policies Bred Collective Insecurity," op. cit.

17. Paul A. Anderson and Stuart J. Thorson, "Artificial Intelligence Based Simulation of Foreign Policy Decision-Making," Behavioral Science 27 (April 1982), pp. 176-193; Dwain Mefford, "Case Studies and Case Study Systems," paper presented to the annual meeting of the International Studies Association, Los Angeles, March 19-22, 1980; also Mefford's "Formulating Foreign Policy on the Basis of Historical Analogies: An Application of Developments in Artificial Intelligence," paper presented to the annual meeting of the International Studies Association, Atlanta, March 27-31, 1984; and his "Changes in Foreign Policy Across Time: The Logical Analysis of a Succession of Decision Problems Using Logic Programming," in Dynamic Models of International Conflict, ed. Urs Luterbacher and Michael D. Ward (Boulder, CO: Lynne Rienner Publishers, 1985); James P. Bennett, "Data Stories: Learning About Learning from the U.S. Experience in Vietnam," in Foreign Policy Decision Making: Perception, Cognition, and Artificial Intelligence, ed. Donald A. Sylvan and Steve Chan (New York: Praeger, 1984), pp. 227-279; Philip A. Schrodt, "Pattern Recognition and the Analysis of International Event Sequences in the Middle East," paper presented to the annual meeting of the International Studies Association, Washington, DC, March 5-9, 1985; and also Schrodt's "Short Term Predictions of International Events Using Pattern Matching," paper presented to the annual meeting of the International Studies Association, Washington, DC, April 14-18, 1987; Scott Bennett and Philip A. Schrodt, "Linear Event Patterns in WEIS Data," paper presented to the annual meeting of the American Political Science Association, Chicago, September 3-6, 1987.

18. McDermott, "Artificial Intelligence Meets Natural Stupidity," op. cit., pp. 156, 157.

APPENDICES

Appendix A: The RICHLIT DBMS

In support of the literature review reported in Chapter Two, I used Ashton-Tate's dBase II package to create a tool called the Richardson Literature Data Base Management System (RICHLIT DBMS). The RICHLIT DBMS consists of a collection of files including the .dbf (database), .prg (program), and .frm (report format), and .ndx (index) file types. There are multiple data tables (or database files, in Ashton-Tate's terminology) in the RICHLIT DBMS, comprised of the following fields: ID CODE, TITLE, AUTHORS, DATE, LOCATOR-SOURCE, ACQUIRED, MODE, TYPE, MATHEMATICAL STRUCTURE, MODEL FORM, ESTIMATOR, DEPENDENT VARIABLES, INDEPENDENT VARIABLES, OPERATIONAL VARIABLES, FOCUS, TECHNIQUE, RESULTS, and QUESTIONS. This tool is not useful beyond its intended purpose, although the concept is valid for application to other literature reviews.

The RICHLIT DBMS greatly facilitated the analysis of this complex body of literature. The data tables contain more than 600 records as of 1989 and continue to grow. Each new citation becomes the nucleus of a new record in the database. As the works are received and abstracted, the structure of each record is progressively filled in. Thus, the RICHLIT DBMS remains incomplete. Nearly one quarter of the records are in only partial states of completion. Nonetheless, it was determined that enough information had been gathered to support the analysis reported in Chapter Two.

The database contains a significant amount of data. The bibliographic citation for each record in the database is listed in the section titled, "Selected Research Bibliography of the Richardson Tradition." Considering the length of the typed bibliography of this book and that these citations occupy only three of the eighteen fields for each record in the database, one can appreciate the amount of information contained in the complete database. This data is available to researchers upon request to the author.

Appendix B: Documentation on ARES System

This appendix presents technical and informational documentation on ARES. This includes notes on modifying several aspects of the system, details of the computing environment, compatibility information, error recovery, and sample output. Actual listings of the code are to be found in Appendix C.

USER-ORIENTED QUALITIES

The "flexibility" of a simulation system refers to the extent to which it is intelligible and modifiable. The high degree of intelligibility of the higher level LISP implementations--much touted in earlier sections of this book--is for the most part not realized in ARES. It probably cannot be said that the functions comprising ARES as a whole are any more intelligible to the casual observer than would be a system of simultaneous equations. However, some of the theoretically critical sections of the code--especially the conditionals in the SUPPLY and REACT functions--are easy to interpret. The relative transparency of the critical conditionals should be counted as one of the clear advantages of this approach.

The relative intelligibility of the critical conditionals also figures in assessing the modifiability of the system. For any new methodology to be truly useful, it must be easily modified by researchers who are not versed in software development. Otherwise the new approach will end up being as remote to most of the discipline as is the optimal control theory approach. On this score, ARES again counts as a step forward. The system can be modified to reflect new assumptions about the relationships among the various objects and attributes comprising the system. This is essential in any simulation system intended to support counter-factual analysis.

Some of the advanced meta-knowledge capabilities demonstrated in large-scale expert systems include self-reporting and self-tracing, so that a simulation can "explain" how it arrived at certain conclusions. This will have an obvious impact on the ability of simulation users, whether researcher or policy maker, to estimate the reliability of the findings and the risk of acting on those findings. While these capabilities are not implemented in ARES, they remain available for incorporation in future generations of this class of simulation systems.

MODIFYING ARES

A critic may object to ARES with a sharp cry, "Why, this REACT function is utter nonsense, a completely unrealistic set of strategies. A nation would not act thus!" In such an event, it is easy to alter the assumptions embedded within ARES that control the action-reaction process. This is done by changing the values in the SUPPLY and REACT functions. The allocative priorities associated with each of the five strategies implemented for ARES are readily evident in an inspection of the SUPPLY function. These allocative priorities can be replaced in order to implement new assumptions about the relationships between strategy and allocation of defense expenditures.

Likewise, the reaction responses of MELOS can be readily reconfigured to reflect alternate assumptions about the strategic behavior of city-states in armaments accumulation processes. By simply rearranging the names of the strategies that MELOS puts into place under different environmental conditions, one can completely change the dynamics of the ARES environment. In order to implement entirely new strategies, it is only necessary to add a new reaction scenario to the REACT function and to alter the SUPPLY function in order to associate the new strategy with a set of allocative priorities.

COMPUTING ENVIRONMENT

ARES was created in 1983 and 1984 on an IBM XT with 256K RAM running under PC-DOS 2.1. It is written in IQLISP 1.4. ARES will run under DOS 2.1 with as little memory as 144K, and though it will run more slowly, it can also be supported under PC-DOS 1.1 with only 64K. The highly simplified nature of the simulation model means that processing time with a smaller amount of internal memory presents no great problem. On the XT system, ARES boots from a hard disk in 12 seconds, and about 45 seconds from a floppy. Processing time depends upon the length of the data time-series, at approximately 2.4 seconds per period (assuming a system speed of 4.77 Mhz). A fifty-year time-series thus runs in about two minutes, with 95% of this time consumed by the graphics output function.

ARES has been tested under all releases of MS-DOS from 1.1 up to 4.0, on a variety of IBM, Compaq, and clone machines using 8088, 8086, 80286 and 80386 architectures. The Integral Quality Lisp interpreter operated as expected on all true compatibles tested. Weird things can happen in the IBM MCA architecture. Smooth execution on the faster standard machines--particularly the new machines executing in the 16-25 Mhz range--requires modifying the PAUSE functions to give appropriate delays for reading the screens. This is most easily done by adding several more declarations of the PAUSE function. ARES has not been tested under OS/2, but it does run under Windows/386.

ARES Version 1.0 is configured to run on systems with a monochrome monitor; ARES Version 1.1 is the color version. Version 1.1 will run on a monochrome monitor without degrading substantive performance characteristics, but also entirely without graphics support. Conceptually, Versions 1.0 and 1.1 are identical, although Version 1.0 uses a table rather than a graph to display run-time results, allowing it to run on machines without a graphics adapter.

ERROR RECOVERY

In most instances bad data will cause ARES to print an error message and continue processing. This will yield bad results (GIGO) but should not cause the system to crash or hang. However, in Version 1.1, a defense budget of zero will causes a fatal error in the graphics routine, invoking the IQLISP error monitor. Recovery is possible by entering "^^" (less the quotes) to reset to top level. Version 1.0 is required if the user wishes to experiment with an inaction/inaction hypotheses involving defense budgets with a value of zero.

SAMPLE OUTPUT

The REPORT function creates an output file object called ARESDATA.OUT in the course of each simulation run. If ARESDATA.OUT already exists, it is renamed ARESDATA.BAK and the new version of the file is given the original name. If ARESDATA.BAK already exists, it is deleted. Thus, the user may make two runs in the ARES simulation system before it is necessary to leave the simulation environment in order to rename the output files before continuing with the experimental series. The format of ARESDATA.OUT is as follows:

```
THE STATUS OF DELOS IN PERIOD 6 WAS: ((NAVY . 16)
(ARMY . 11) (AIR . 26) (STRATEGY . POWER))

AND THE STATUS OF MELOS IN PERIOD 6 WAS: ((NAVY . 11)
(ARMY . 16) (AIR . 26) (STRATEGY . BATTLE))
```

Appendix C: Documentation on ARES Code

```
~~
~~GREET.ARS
~~
~~PRINTS A GREETING UPON ENTRY INTO ARES ENVIRONMENT
~~CALLED BY TOPLEVEL.ARS
~~CALLS PAUSE5.ARS AND RETURNS A CALL TO MENU.ARS
~~
(LAMBDA () (PROG () (DISPLAYMODE 640)(COLORMODE 0 4)
   (WINCLR)(WINCURS '(5 32))
   (PRINC "WELCOME TO ARES")(WINCURS '(7 30))
   (PRINC "Arms RacE Simulation")(WINCURS '(10 34))
   (PRINC "Version 1.1")(WINCURS '(16 26))
   (PRINC "A SIMULATION OF ARMS RACES")(WINCURS '(17 28))
   (PRINC "BETWEEN DELOS AND MELOS")(WINCURS '(22 21))
   (PRINC (LIST 'THIS 'RUN 'WAS 'PERFORMED 'ON (DATE)))
      (LINE 1 1 1 190 1)(LINE 1 1 630 1 1)
      (LINE 1 190 630 190 1)(LINE 630 190 630 1 1)
   (PAUSE5)
(RETURN (MENU))]

~~
~~TOPLEVEL.ARS
~~
~~REDEFINES TOPLEVEL OF IQLISP ENVIRONMENT
~~TO BOOT ARES GREETING ROUTINE
~~CALLED BY USER FROM SYSTEM
~~WITH THE COMMAND "IQ ARES"
~~ALSO CALLED BY MENU.ARS TO RESET
~~THE DEFAULT INITIAL CONDITIONS
~~CALLS GREET.ARS
~~
(LAMBDA (RESET)
   (COND (RESET (SETQ ELVL 0)))
   (SAVE "ARES")
(GREET)]

~~
~~OPERATE.ARS
~~
~~THIS FUNCTION INSTANTIATES STRATEGY FOR DELOS;
~~FIRST IT TAKES THE INPUT FOR THE YEAR AND THEN
~~IT FINDS THE CURRENT STRATEGY AND CALLS SUPPLY.
~~CALLED BY REACT.ARS
~~CALLS SUPPLY.ARS
~~
(LAMBDA (YEAR) (PROG ()
   (PUT 'DELOS (CAR YEAR) 'STRATEGY)
   (PUT 'DELOS (CADR YEAR) 'BUDGET)
   (SUPPLY 'DELOS)]
```

```
~~
~~MENU.ARS
~~
~~CONTROLS ARES MAIN MENU OF CHOICES
~~CALLED BY MENU
~~CALLS DATA.ARS, INITIAL.ARS, DRIVER.ARS,
~~  REPORT.ARS, TOPLEVEL.ARS, MENU.ARS, AND (EXIT)
~~
(LAMBDA () (PROG (CHOICE)
  (WINCLR)(DISPLAYMODE 640)(COLORMODE 0 2)(WINCURS '(1 30))
  (PRINC "ARES MAIN MENU") (WINCURS '(4 16))
  (PRINC "1. Input Data for the Current Run")
    (WINCURS '(6 16))
  (PRINC "2. Set Initial Conditions for Current Run")
    (WINCURS '(8 16))
  (PRINC "3. Execute the Current Run")(WINCURS '(10 16))
  (PRINC "4. Display Results of the Previous Run")
    (WINCURS '(12 16))
  (PRINC "5. Reset ARES Simulation System")
    (WINCURS '(14 16))
  (PRINC "6. Exit ARES to Disk Operating System")
    (WINCURS '(16 16))(WINCURS '(18 17))
  (PRINC "PLEASE SELECT A NUMBER AND ENTER HERE ===>")
    (LINE 1 1 1 160 1)(LINE 1 1 630 1 1)
    (LINE 1 160 630 160 1)(LINE 630 160 630 1 1)
    (LINE 1 130 630 130 1)(LINE 1 20 630 20 1)
    (WINCURS '(18 61))
  (SETQ CHOICE (READ))
  (COND ((EQ CHOICE 1)(DATA))
      ((EQ CHOICE 2)(INITIAL))
      ((EQ CHOICE 3)(DRIVER))
      ((EQ CHOICE 4)(REPORT))
      ((EQ CHOICE 5)(TOPLEVEL))
      ((EQ CHOICE 6)(EXIT))
    (T (RETURN (MENU)))]
```

```
~~
~~READIN.ARS
~~
~~READS ONE CHARACTER AT A TIME FROM KEYBOARD
~~UNTIL A CARRIAGE RETURN IS ENCOUNTERED;
~~THEN IT PUTS INPUT ON PROPERTY LIST OF COUNTRY
~~CALLED BY STORE.ARS
~~CALLS READIN.ARS AND APPEND.ARS
~~
(LAMBDA (Y W)
   (COND ((EQUAL Y "ER")
         (PUT COUNTRY (READLIST W) PROPERTY))
   (T (COND ((NULL W)
              (SETQ W (LIST Y)))
         (T (SETQ W (APPEND W (LIST Y)))))
              (SETQ Y (TYI))
      (COND ((NEQUAL Y "ER")
              (PRINC Y)))
(READIN Y W)))]
```

```
~~
~~DATA.ARS
~~
~~PROMPTS THE USER TO INPUT DATA FOR THE RUN
~~CONSTRUCTS THE HISTORY SCRIPT OF THE RUN
~~
(LAMBDA ()   (PROG (PERIOD LENGTH D M)
   (WINCLR)
   (PRINT '(HOW MANY YEARS OF DATA IN THIS RUN?))
   (SETQ LENGTH (READ)) (SETQ CHUNK NIL)
   (SETQ PERIOD 1)(WINCLR)
   (PUT 'MELOS '(0 0 0) 'XPOS)
   (PUT 'DELOS '(0 0 0) 'XPOS) LOOP
   (COND ((GT PERIOD LENGTH)(MENU)))
   (PRINT (LIST 'PLEASE 'ENTER 'STRATEGY 'FOR 'DELOS
      'IN 'PERIOD PERIOD)) (SETQ HUNK (READ))
   (PRINT (LIST 'PLEASE 'ENTER 'BUDGET 'FOR 'DELOS
      'IN 'PERIOD PERIOD)) (SETQ D (READ))
   (PRINT (LIST 'PLEASE 'ENTER 'BUDGET 'FOR 'MELOS
      'IN 'PERIOD PERIOD)) (SETQ M (READ))
   (SETQ HUNK (LIST HUNK D M))
   (COND ((NULL CHUNK)
         (SETQ CHUNK (LIST HUNK)))
   (T (SETQ CHUNK (APPEND CHUNK (LIST HUNK)))))
   (SETQ HISTORY CHUNK) (SETQ PERIOD (ADD1 PERIOD))
(WINCLR) (GO LOOP)]

~~
~~INITIAL.ARS
~~
~~
~~THIS ROUTINE SETS INITIAL CONDITIONS FOR SIMULATION RUN
~~THE INNER LOOP CONTROLS INPUT OF PROPERTIES
~~THE OUTER LOOP CONTROLS INPUT OF COUNTRIES
~~CALLED BY DRIVER.ARS
~~CALLS STORE.ARS AND FIND.ARS
~~
(LAMBDA () (PROG (COUNT)
   (SETQ WORLD '(DELOS MELOS ROW))
   (SETQ COUNT 0) LOOP2 (WINCLR)
   (SETQ ITEMS '(STRATEGY AIR ARMY NAVY))
   (SETQ COUNT (+ COUNT 1))
   (SETQ COUNTRY (CAR WORLD))
   (SETQ WORLD (CDR WORLD)) LOOP1
(COND ((EQUAL COUNT 3)
      (RETURN (MENU)))
   ((EQUAL ITEMS NIL)(GO LOOP2)))
   (SETQ PROPERTY (CAR ITEMS))
   (SETQ ITEMS (CDR ITEMS))
   (PRINT (LIST 'INITIAL PROPERTY 'FOR COUNTRY 'IS?===>))
   (STORE COUNTRY PROPERTY)
   (PRINT (LIST COUNTRY 'INITIAL PROPERTY 'IS
(FIND COUNTRY PROPERTY))) (GO LOOP1)]
```

```
~~
~~FIND.ARS
~~
~~FINDS AND RETURNS A SPECIFIED PROPERTY
~~CALLED BY INITIAL.ARS
~~
(LAMBDA (X Y) (PROG (Z)
   (COND ((NUMBERP X) (RETURN 'NUMBER))
         ((NOT (ATOM X)) (RETURN 'NOT-ATOM))
   (T (AND (SETQ X (PROPS X)) (SETQ Z (CAR X)))))
LOOP
   (COND ((OR (NULL (CAR Z))(EQUAL Y (CAR Z)))
         (RETURN (CDR Z)))
   (T (AND (SETQ X (CDR X))
      (SETQ Z (CAR X))
(GO LOOP))))]

~~
~~DRIVER.ARS
~~
~~THIS FUNCTION DRIVES THE SIMULATION;
~~IT LOOPS THRU A TEST-OPERATE-REACT CYCLE,
~~AND OUTPUTS A REPORT AT THE END OF THE DATA.
~~CALLED BY MENU.ARS
~~CALLS REACT.ARS, OPERATE.ARS, AND REPORT.ARS
~~
(LAMBDA ()    (PROG ()
   (SETQ HISTORY CHUNK)
TIME
   (SETQ YEAR (CAR CHUNK))
   (COND ((EQ CHUNK NIL)(RETURN (REPORT))))
      (OPERATE YEAR)
      (REACT  YEAR)
      (SETQ CHUNK (CDR CHUNK))
(GO TIME)]

~~
~~STORE.ARS
~~
~~PUTS A VALUE READ IN FROM THE TERMINAL
~~ONTO THE PROPERTY LIST OF THE COUNTRY
~~CALLED BY INITIAL.ARS
~~CALLS READIN.ARS
~~
(LAMBDA (COUNTRY PROPERTY Y W)
   (PRINT (READLIST (EXPLODEC (CONCAT PROPERTY "?"))))
      (CLRBFI) (SETQ Y (TYI))
   (COND ((EQUAL Y "ER")
         (PUT COUNTRY 'VOID PROPERTY))
   (T (PRINC Y)
      (READIN Y W)))
   (PRINT (LIST PROPERTY (GET COUNTRY PROPERTY)))]
```

```
~~
~~REACT.ARS
~~
~~THIS FUNCTION CONTROLS CHANGE IN MELOS;
~~FIRST THE ANNUAL MILITARY BUDGET FOR MELOS IS SET;
~~THEN MELOS INFERS FROM
~~DELOS' CAPABILITIES THE STRATEGY OF DELOS, THEREBY
~~SELECTING AN APPROPRIATE COUNTER-STRATEGY FOR MELOS.
~~CALLED BY DRIVER.ARS
~~CALLS OPERATE.ARS AND SUPPLY.ARS
~~
(LAMBDA (YEAR BUDGET STRATEGY) (PROG ()
   (PUT 'MELOS (CADDR YEAR) 'BUDGET)
(COND
~~
~~THE NEXT FIFTEEN LINES INFER THE DELOS ORDER OF BATTLE
~~AND IDENTIFY A COUNTER-STRATEGY FOR MELOS
~~
   ((AND (GE (GET 'DELOS 'AIR)(GET 'DELOS 'NAVY))
      (GE (GET 'DELOS 'NAVY)(GET 'DELOS 'ARMY)))
             (PUT 'MELOS 'BATTLE 'STRATEGY))
   ((AND (GE (GET 'DELOS 'NAVY)(GET 'DELOS 'ARMY))
      (GE (GET 'DELOS 'ARMY)(GET 'DELOS 'AIR)))
             (PUT 'MELOS 'DEFENSE 'STRATEGY))
   ((AND(GE (GET 'DELOS 'ARMY)(GET 'DELOS 'AIR))
      (GE (GET 'DELOS 'AIR)(GET 'DELOS 'NAVY)))
             (PUT 'MELOS 'CONTROL 'STRATEGY))
   ((AND (GE (GET 'DELOS 'AIR)(GET 'DELOS 'ARMY))
      (GE (GET 'DELOS 'ARMY)(GET 'DELOS 'NAVY)))
             (PUT 'MELOS 'INVADE 'STRATEGY))
   ((AND (GE (GET 'DELOS 'NAVY)(GET 'DELOS 'AIR))
      (GE (GET 'DELOS 'AIR)(GET 'DELOS 'ARMY)))
             (PUT 'MELOS 'POWER 'STRATEGY))
~~
~~THE NEXT FEW LINES ACCOUNT FOR THE ORDERS OF BATTLE WHICH
~~DO NOT FIT THE PROFILE OF ANY OF THE PRINCIPAL MISSIONS;
~~AND FINALLY THE MELOS COUNTER-STRATEGY IS INSTANTIATED
~~
   ((AND (GT (GET 'DELOS 'AIR)(GET 'DELOS 'NAVY))
      (GT (GET 'DELOS 'AIR)(GET 'DELOS 'ARMY)))
         (PUT 'MELOS 'BATTLE 'STRATEGY))
   ((AND (GT (GET 'DELOS 'NAVY)(GET 'DELOS 'AIR))
      (GT (GET 'DELOS 'NAVY)(GET 'DELOS 'ARMY)))
         (PUT 'MELOS 'DEFENSE 'STRATEGY))
   ((AND (GT (GET 'DELOS 'ARMY)(GET 'DELOS 'AIR))
      (GT (GET 'DELOS 'ARMY)(GET 'DELOS 'NAVY)))
         (PUT 'MELOS 'CONTROL 'STRATEGY)))
(SUPPLY 'MELOS)]
```

```
~~
~~SUPPLY.ARS
~~
~~THIS FUNCTIONS CONTROLS RESOURCE ALLOCATION;
~~IT IS USED BY BOTH MELOS AND DELOS
~~TO ALLOCATE THE BUDGET AMONG THE MILITARY ARMS;
~~IT IDENTIFIES THE STRATEGY, AND THEN IT
~~IMPLEMENTS THE APPROPRIATE BUDGET MODEL.
~~CALLED BY OPERATE.ARS AND REACT.ARS
~~
(LAMBDA (COUNTRY)  (PROG ()
   (SETQ OLDAIR (GET COUNTRY 'AIR))
   (SETQ OLDARMY (GET COUNTRY 'ARMY))
   (SETQ OLDNAVY (GET COUNTRY 'NAVY))
   (SETQ OLDLVLS (LIST OLDAIR OLDARMY OLDNAVY))
   (COND ((EQUAL (GET COUNTRY 'STRATEGY) 'POWER)
         (AND (PUT COUNTRY (+ (GET  COUNTRY 'AIR)
            (* (GET 'DELOS 'BUDGET) 5)) 'AIR)
         (AND (PUT COUNTRY (+ (GET COUNTRY 'NAVY)
            (* (GET COUNTRY 'BUDGET) 3)) 'NAVY)
            (PUT COUNTRY (+ (GET COUNTRY 'ARMY)
            (* (GET COUNTRY 'BUDGET) 2)) 'ARMY))))
   ((EQUAL (GET COUNTRY 'STRATEGY) 'INVADE)
      (AND (PUT COUNTRY (+ (GET COUNTRY 'AIR)
         (* (GET COUNTRY 'BUDGET) 2)) 'AIR)
      (AND (PUT COUNTRY (+ (GET COUNTRY 'NAVY)
         (* (GET COUNTRY 'BUDGET) 5)) 'NAVY)
         (PUT COUNTRY (+ (GET COUNTRY 'ARMY)
         (* (GET COUNTRY 'BUDGET) 3)) 'ARMY))))
   ((EQUAL (GET COUNTRY 'STRATEGY) 'DEFENSE)
      (AND (PUT COUNTRY (+ (GET COUNTRY 'AIR)
         (* (GET COUNTRY 'BUDGET) 3)) 'AIR)
      (AND (PUT COUNTRY (+ (GET COUNTRY 'NAVY)
         (* (GET COUNTRY 'BUDGET) 2)) 'NAVY)
         (PUT COUNTRY (+ (GET COUNTRY 'ARMY)
         (* (GET COUNTRY 'BUDGET) 5)) 'ARMY))))
   ((EQUAL (GET COUNTRY 'STRATEGY) 'CONTROL)
      (AND (PUT COUNTRY (+ (GET COUNTRY 'AIR)
         (* (GET COUNTRY 'BUDGET) 3)) 'AIR)
      (AND (PUT COUNTRY (+ (GET COUNTRY 'NAVY)
         (* (GET COUNTRY 'BUDGET) 5)) 'NAVY)
         (PUT COUNTRY (+ (GET COUNTRY 'ARMY)
         (* (GET COUNTRY 'BUDGET) 2)) 'ARMY))))
   ((EQUAL (GET COUNTRY 'STRATEGY) 'BATTLE)
      (AND (PUT COUNTRY (+ (GET COUNTRY 'AIR)
         (* (GET COUNTRY 'BUDGET) 5)) 'AIR)
      (AND (PUT COUNTRY (+ (GET COUNTRY 'NAVY)
         (* (GET COUNTRY 'BUDGET) 2)) 'NAVY)
         (PUT COUNTRY (+ (GET COUNTRY 'ARMY)
         (* (GET COUNTRY 'BUDGET) 3)) 'ARMY))))
   (T (PRINT '(ERROR: BAD STRATEGY))))
(GRAPH COUNTRY OLDLVLS)]
```

```
~~GRAPH.ARS
~~
~~THIS FUNCTION PROVIDES REALTIME GRAPHICS OUTPUT
~~TO THE SCREEN FOR DISPLAY OF FORCE POSTURE CHANGES
~~OVER THE COURSE OF THE SIMULATION RUN; THIS ALLOWS
~~THE USER TO OBSERVE THE DYNAMICS OF THE SYSTEM AS
~~THEY UNFOLD DURING THE RUN.
~~LOOP1 SETS THE Y-COORDINATE AND PAINTS THE PIXEL;
~~LOOP2 SETS THE X-COORDINATE AND RESETS Y;
~~LOOP3 SETS SERVICE, WINDOW & INCREMENT, AND RESETS X.
~~THE FIRST CONDITIONAL SETS THE COUNTRY.
~~CALLED BY SUPPLY.ARS
~~
(LAMBDA (COUNTRY OLDLVLS)  (PROG (X Y)
  (DISPLAYMODE 321) (COLORMODE 0 4) (COLORMODE 1 0)
  (SETQ ARMS '(AIR ARMY NAVY)) (SETQ NEWPOS NIL)
  (SETQ XPOS (GET COUNTRY 'XPOS))
  (SETQ WIND1  (WINDOW '(0 0) '(4 80)))
  (SETQ WIND2  (WINDOW '(4 0) '(4 80)))
  (SETQ WIND3  (WINDOW '(8 0) '(4 80)))
  (SETQ WIND4 (WINDOW '(12 0) '(4 80)))
  (SETQ WIND5 (WINDOW '(17 0) '(4 80)))
  (SETQ WIND6 (WINDOW '(21 0) '(4 80)))
  (PRINC "DELOS" WIND1) (WINCURS '(2 0) WIND1)
  (PRINC "      AIR FORCE" WIND1) (WINCURS '(2 0) WIND2)
  (PRINC "      ARMY" WIND2) (WINCURS '(2 0) WIND3)
  (PRINC "      NAVY" WIND3) (WINCURS '(0 0) WIND4)
  (PRINC "MELOS" WIND4) (WINCURS '(2 0) WIND4)
  (PRINC "      AIR FORCE" WIND4) (WINCURS '(1 0) WIND5)
  (PRINC "      ARMY" WIND5) (WINCURS '(1 0) WIND6)
  (PRINC "      NAVY" WIND6)
  (COND ((EQUAL COUNTRY 'DELOS)(AND (SETQ CYPOS '(8 40 72))
          (SETQ WINDLIST '(WIND1 WIND2 WIND3))))
  (T (AND (SETQ CYPOS '(104 136 168))
        (SETQ WINDLIST '(WIND4 WIND4 WIND6)))))
LOOP3
  (COND ((EQUAL OLDLVLS NIL)
    (AND (PUT COUNTRY NEWPOS 'XPOS)(RETURN))))
      (SETQ OLDLEVEL (CAR OLDLVLS))(SETQ OLDLVLS
        (CDR OLDLVLS))
      (SETQ YPOS (CAR CYPOS))(SETQ CYPOS (CDR CYPOS))
      (SETQ WIND (CAR WINDLIST))(SETQ WINDLIST
        (CDR WINDLIST))
      (SETQ ARM (CAR ARMS))(SETQ ARMS (CDR ARMS))
      (SETQ I (/ (- (GET COUNTRY ARM) OLDLEVEL)
      (GET COUNTRY 'BUDGET)))
      (SETQ X (CAR XPOS))(SETQ XPOS (CDR XPOS))(SETQ INDEX X)
LOOP2
  (COND ((EQ X (+ INDEX I))
      (AND (COND ((NULL NEWPOS)(SETQ NEWPOS (LIST X)))
  (T (SETQ NEWPOS (APPEND NEWPOS (LIST X))))) (GO LOOP3))))
  (SETQ X (ADD1 X)) (SETQ Y YPOS)
LOOP1
  (COND ((GT Y (+ YPOS 6))(GO LOOP2)))
      (POINT X Y '3 WIND)(SETQ Y (ADD1 Y))
(GO LOOP1)]
```

```
~~
~~REPORT.ARS
~~
~~SIMPLIFIED REPORT PRINTER
~~
(LAMBDA ()    (PROG ()    (PAUSE5)
  (WINCLR)(DISPLAYMODE 640)(COLORMODE 0 1)(WINCURS '(1 17))
  (PRINC "THE RESULTS OF THE RUN ARE AS FOLLOWS:")
    (WINCURS '(4 5))
  (PRINC "THE POSTURE OF DELOS AT THE END WAS:")
  (TERPRI)(TERPRI)(PRINC (CDR (PROPS 'DELOS))) (TERPRI)
    (WINCURS '(9 5))
  (PRINC "THE POSTURE OF MELOS AT THE END WAS:")
  (TERPRI)(TERPRI)(PRINC (CDR (PROPS 'MELOS)))(TERPRI)
  (WINCURS '(14 5))(PRINC "THE HISTORY OF THE SYSTEM WAS:")
  (TERPRI)(PRINT (EVAL 'HISTORY))
    (LINE 0 1 0 160 1)(LINE 0 1 630 1 1)
    (LINE 0 160 630 160 1)(LINE 630 160 630 1 1)
    (LINE 0 20 630 20 1)(LINE 0 65 630 65 2)
    (LINE 0 105 630 105 2) (PRINTC
    (LIST "THE STATUS OF DELOS IN PERIOD  " PERIOD "WAS:"
    (CDR (PROPS 'DELOS))
    "AND THE STATUS OF MELOS IN PERIOD "  PERIOD "WAS:"
    (CDR (PROPS 'MELOS)))
  (OUTPUT "ARESDATA.OUT"))
  (PAUSE5)(PAUSE5)
(RETURN (MENU))]


~~
~~APPEND.ARS
~~
~~ASSEMBLES EXPRESSIONS ONE CHARACTER AT A TIME
~~FROM KEYBOARD INPUT
~~CALLED BY READIN.ARS AND DATA.ARS
~~CALLS APPEND.ARS
~~
(LAMBDA (L1 L2)
   (COND ((ATOM L1) L2)
        (T (CONS (CAR L1)
(APPEND (CDR L1) L2)))))


~~
~~ATOMCAR.ARS
~~
(LAMBDA (X)
(READLIST
(LIST
(CAR (EXPLODEC X)))))
```

```
~~
~~ATOMCDR.ARS
~~
(LAMBDA (X)
  (READLIST
(CDR (EXPLODEC X))))


~~
~~PAUSE2.ARS
~~
~~CAUSES A 2 SECOND SUSPENSION OF THE PROGRAM
~~BY ASKING THE COMPUTER TO COUNT TO 500
~~
(LAMBDA ()  (PROG (X)
  (SETQ X 1)
LOOP
  (COND ((EQUAL X 500)
       (RETURN)))
  (SETQ X (+ X 1))
(GO LOOP)]


~~
~~PAUSE5.ARS
~~
~~CAUSES A 5 SECOND SUSPENSION OF THE PROGRAM
~~BY ASKING THE COMPUTER TO COUNT TO 1250
~~
(LAMBDA () (PROG (X)
  (SETQ X 1)
LOOP
  (COND ((EQUAL X 1250)
       (RETURN)))
  (SETQ X (+ X 1))
(GO LOOP)]
```

SELECTED BIBLIOGRAPHY

The Richardson Literature

Abelson, R. "A 'Derivation' of Richardson's Equations." Journal of Conflict Resolution. 7:1 (March 1963): 13-15.

Abolfathi, Farid. "Defense Expenditures in the Persian Gulf: Internal, Interstate, and International Factors in the Iraqi-Iranian Arms Race, 1950 to 1969." In Exploring Competitive Armaments Processes, pp. 99-131. Edited by W.L. Hollist. New York: Marcel Dekker, 1978.

Abt, Clark C., and Gorden, Morton. "Report on Project Temper." In Theory and Research on the Causes of War, pp. 245-262. Edited by D.G. Pruitt and R.C. Snyder. Englewood Cliffs, NJ: Prentice-Hall, 1969.

Achen, Christopher. "A Misperception Model for a Spiral Arms Race." Paper presented to the annual meeting of the International Studies Association, Atlanta, March 27-31, 1984.

Afheldt, H., and Sonntag, P. "Stability and Deterrence Through Strategic Nuclear Arms." Journal of Peace Research. 3 (1973): 245-250.

Albrecht, U. "The Costs of Armamentism." Journal of Peace Research. 3 (1973): 265-283.

Alcock, Norman Z. "Arms Races--A Cause of War?" Peace Research. 2:2 (1970): 3.

---. "An Empirical Measure of International Threat: Some Preliminary Implications for the Middle East Conflict." Peace Research Society (International) Papers. 15 (1970): 51-72.

---. The War Disease. Oakville, Ontario: CPRI Press, 1972.

Alcock, N.Z., and Lowe, Keith W. "The Vietnam War as a Richardson Process." Journal of Peace Research. 6:2 (April 1969): 105-112.

Alker, Hayward R., Jr. "The Structure of Social Action in an Arms Race." Paper presented to the North American Peace Research Society (International) meeting, Cambridge, MA, November 10-12, 1968.

---. "Research Paradigms and Mathematical Politics." Paper presented to the International Political Science Association meeting, Mannheim, Germany, 1971.

---. "Global Modeling Alternatives: The First Twenty Years." In Theories, Models, and Simulations in International Relations, pp. 9-37. Edited by Michael D. Ward. Boulder, CO: Westview Press, 1985.

Alker, Hayward R., Jr., and Brunner, R.D. "Simulating International Conflict: A Comparison of Three Approaches." International Studies Quarterly. 13 (1969): 70-110.

Alker, Hayward R., Jr.; Deutsch, Karl W.; and Stoetzel, Antoine H., eds. Mathematical Approaches to Politics. San Francisco: Jossey-Bass, 1973.

Allan, Pierre. "The Impact of International Climate on Arms Races." Paper presented to the annual meeting of the Western Section of the Peace Science Society International, Stanford, CA, February 26-27, 1979.

---. "Determinants of Swiss Military Expenditures." In Small Countries and International Structural Adjustment, pp. 171-181. Edited by I. Dobozi, C. Keller, and H. Matejka. Geneva: IUHEI, 1982.

---. Crisis Bargaining and the Arms Race: A Theoretical Model. Cambridge, MA: Ballinger Publishing Co., 1983.

---. "Towards Building Social Clocks." Paper presented to the conference on Dynamic Models of Conflict, Boulder, CO, October 31 to November 3, 1984.

Allan, Pierre, and Luterbacher, Urs. "The Future of East-West Relations: A Computer Simulation of Five Scenarios." In East-West Relations, Vol. 2, pp. 285-318. Edited by Daniel Frei and Dieter Ruloff. Cambridge, MA: Oelgeschlager, Gunn, and Hain, 1983.

---. "Modelling East-West Strategic Relations in the 1980's." In East-West Relations in the Mid-Eighties, pp. 35-67. Edited by I. Dobozi and H. Matejka. Budapest: 1984.

Allison, Graham T. "Questions about the Arms Race: Who's Racing Whom? A Bureaucratic Perspective." In American Defense Policy, 4th ed., pp. 424-441. Edited by John E. Endicott and Roy W. Stafford. Baltimore: Johns Hopkins Press, 1977.

Allison, Graham T., and Morris, F.A. "Armaments and Arms Control: Exploring the Determinants of Military Weapons." Daedalus. 104:3 (Summer 1975): 99-129.

Altfeld, Michael F. "Arms Races?--And Escalation? A Comment on Wallace." International Studies Quarterly. 27:2 (June 1983): 225-231.

Altfeld, Michael F., and Morrow, James. "A Microeconomic Model of Military Expenditures." Paper presented to the annual meeting of the International Studies Association, Atlanta, March 27-31, 1984.

Anderton, Charles F. "Arms Race Modeling: Categorization and Systematic Analysis." Paper presented to the annual meeting of the International Studies Association, Washington, DC, March 5-9, 1985.

Anscombe, F.J. et al. Applications of Statistical Methodology to Arms Control and Disarmament. Princeton, NJ: Mathematica, 1963.

Anselin, Luc. "A Simulation Framework for Modeling Dynamics in Policy Space." Conflict Management and Peace Science. 7:1 (Fall 1983): 25-38.

Ashley, Richard K. The Political Economy of War and Peace: The Sino-Soviet-American Triangle and the Modern Security Problematique. New York and London: Nichols and Pinter Publishers, 1980.

---. "The Eye of Power: The Politics of World Modeling." International Organization. 37:3 (Summer 1983): 495-535.

Azar, Edward E. "The Dimensionality of Violent Conflict: A Quantitative Analysis." Peace Research Society (International) Papers. 15 (1970): 122-167.

---. "The Theory of Protracted Social Conflict and the Challenge of Transforming Conflict Situations." In Conflict Processes and the Breakdown of International Systems, pp. 81-99. Edited by Dina A. Zinnes. Denver: Graduate School for International Studies, 1983.

Azar, Edward E.; Bennett, James P.; and Sloan, Thomas J. "Steps Toward Forecasting International Interactions." Peace Research Society (International) Papers. 23 (1970): 27-67.

Bacskai, Thomas. "The Armaments Race and the Developing Countries." Peace Research Society Papers. 6 (1966): 31-35.

Banks, F.E. "The Equilibrium Theory of Arms Races: A Comment." Journal of Peace Research. 12 (1975): 235-237.

Bateson, Gregory. "The Pattern of an Armaments Race--Part I: An Anthropological Approach." Bulletin of the Atomic Scientists. 2:5/6 (1946): 10-11.

---. "The Pattern of an Armaments Race--Part II: An Analysis of Nationalism." Bulletin of the Atomic Scientists. 2:7/8 (1946): 26-28.

Baugh, William H. III. "Systemic Military Stability Among Nation-States: An Operations Analysis." Ph.D. dissertation, Indiana University, 1973.

---. "An Operations Analysis Model for the Study of Nuclear Missile System Policies." In Mathematical Models in International Relations, pp. 274-305. Edited by Dina A. Zinnes and John V. Gillespie. New York: Praeger, 1976.

---. "Is There an Arms Race: Alternative Models of Slow Growth Processes." General Systems. 22 (1977): 445-450.

---. "Transient-Response Analysis of Richardson Type Arms Race Models." In Mathematical Systems in International Relations Research, pp. 221-248. Edited by John V. Gillespie and Dina A. Zinnes. New York: Praeger, 1977.

---. "Major Powers and Weak Allies: Stability and Structure in Arms Race Models." Journal of Peace Science. 3:1 (Spring 1978): 45-54.

---. "Arms Transfers and War Among Third World States: A Poisson Regression Study." Paper presented to the annual meeting of the International Studies Association, Mexico City, April 5-9, 1983.

Baugh, William III, and Squires, Michael. "Arms Transfers and the Onset of War, Part I: Scalogram Analysis of Transfer Patterns." Paper presented to the annual meeting of the International Studies Association, Los Angeles, March 19-22, 1980.

---. "Arms Transfers and the Onset of War, Part II: Wars in Third World States, 1950-1965." Paper presented to the annual meeting of the International Studies Association, Philadelphia, 1981.

Bennett, James P., and Alker, Hayward R., Jr. "When Collective Security Policies Bred Collective Insecurity: The War of the Pacific in a World Politics Simulation." In Problems of World Modeling, pp. 215-302. Edited by Karl W. Deutsch, Bruno Fritsch, Helio Jaguaribe, and Andrei Markovits. Cambridge, MA: Ballinger Publishing Co., 1977.

Bernd, Joseph L., ed. Mathematical Applications in Political Science, Vol. 2. Dallas: Southern Methodist University, 1966.

Bird, Charles. "An Infinite Player Cooperative Game Theory Model for the Analysis of the Effect of Various Factions on International Policy Making." In Mathematical Systems in International Relations Research, pp. 396-420. Edited by John V. Gillespie and Dina A. Zinnes. New York: Praeger, 1977.

Bird, Charles, and Sampson, Martin W. III. "A Game Theory Model of OPEC, Oil Consumers, and Oil Companies, Emphasizing Coalition Formations." In Mathematical Models in International Relations, pp. 376-397. Edited by Dina A. Zinnes and John V. Gillespie. New York: Praeger, 1976.

Blankenship, Gilmer L. "An Input-Output Model of Human Population Dynamics." In Mathematical Systems in International Relations Research, pp. 98-120. Edited by John V. Gillespie and Dina A. Zinnes. New York: Praeger, 1977.

Blumberg, A.A. "Model for a Two-Adversary Arms Race." Nature. 234 (November 19, 1971): 158.

Bonder, S., and Farrell, R., eds. Development of Models for Defense Systems Planning. Report SRL 2147 TR 70-2. Ann Arbor: Systems Research Laboratory, University of Michigan, 1970.

Bottome, Edgar M. The Missile Gap: A Study of the Formulation of Military and Public Policy. Cranbury, NJ: Fairleigh Dickinson University, 1970.

---. The Balance of Terror: A Guide to the Arms Race. Boston: Beacon Press, 1971.

Boulding, Kenneth E. Conflict and Defense: A General Theory. New York: Harper and Row, 1962.

---. "Future Directions in Conflict and Peace Studies." Journal of Conflict Resolution. 22:4 (December 1978): 342-354.

Brams, Stephen J. Game Theory and Politics. New York: Free Press, 1975.

---. Superpower Games: Applying Game Theory to Superpower Conflict. New Haven: Yale University Press, 1985.

Brams, Stephen J.; Davis, Morton D.; and Straffin, Philip D., Jr. "The Geometry of the Arms Race." International Studies Quarterly. 23:4 (December 1979): 567-588.

---. "A Reply to 'Detection and Disarmament.'" International Studies Quarterly. 23:4 (December 1979): 599-600.

Brams, Stephen J., and Kilgour, D. Marc. "Deterrence versus Defense: A Game-Theoretic Model of Star Wars." Paper presented to the annual meeting of the American Political Science Association, Chicago, September 3-6, 1987.

---. "Threat Escalation and Crisis Stability: A Game-Theoretic Analysis." American Political Science Review. 81:3 (September 1987): 833-850.

Bremer, Stuart A. "National and International Systems: A Computer Simulation." Ph.D. dissertation, University of Michigan, 1970.

---. Simulated Worlds: A Computer Model of National Decision-Making. Princeton, NJ: Princeton University Press, 1977.

---. "The GLOBUS Model: History, Structure, and Illustrative Results." In Theories, Models, and Simulations in International Relations, pp. 39-77. Edited by Michael D. Ward. Boulder, CO: Westview Press, 1985.

---. "Understanding Government Decision-Making: A Design for a Computer-Simulated Multi-nation, Multi-sector Model." Paper presented to the annual meeting of the International Studies Association, Anaheim, CA, March 25-29, 1986.

---, ed. The GLOBUS Model: Computer Simulation of Worldwide Political and Economic Developments. Boulder, CO: Westview Press, 1987.

Bremer, Stuart, and Cusack, Thomas R. "Anticipation and Arms Accumulation: A Simulation Study of the Impact of Forecasting on Arms Races." IIVG Discussion Paper 80-106. Berlin: Wissenshaftszentrum Berlin, 1980.

---. "The Urns of War: An Application of Probability Theory to the Genesis of War." IIVG Discussion Paper 80-128. Berlin: Wissenschaftszentrum Berlin, 1980.

Bremer, Stuart A., and Hughes, Barry B. "Reducing East-West Conflict and the North-South Gap." Paper presented to the annual meeting of the American Political Science Association, Chicago, September 3-6, 1987.

Brewer, Garry D., and Shubik, Martin. The War Game: A Critique of Military Problem Solving. Cambridge, MA: Harvard University Press, 1979.

Brito, Dagobert L. "A Dynamic Model of an Armaments Race." International Economic Review. 13:2 (1972): 359-375.

Brito, Dagobert L.; Buoncristiani, A.M.; and Intriligator, Michael D. "A New Approach to the Nash Bargaining Problem." Econometrica. 45 (1977): 1163-1172.

Brito, Dagobert L., and Intriligator, Michael D. "A General Equilibrium Model of the Stability of an Armaments Race." Paper presented to the sixth Conference on Circuits and Systems, Asilomar, CA, 1972.

---. "Some Applications of the Maximum Principle to the Problem of an Armaments Race." In Modeling and Simulation, Vol. 4, pp. 140-144. Edited by William G. Vogt and Marlin H. Mickle. Pittsburgh: Instrument Society of America, 1973.

---. "Uncertainty and the Armaments Race." Annals of Economic and Social Measurement. 3:1 (1974): 279-292.

---. "An N-Country Model of an Armaments Race." N.p.: (mimeo), 1974.

---. "A Fixed Point Approach to Multiagent Adaptive Control." Annals of Economic and Social Measurement. 6 (1977): 137-145.

---. "Nuclear Proliferation and the Armaments Race." Journal of Peace Science. 2 (1977): 231-238.

---. "Strategic Weapons and the Allocation of International Rights." In Mathematical Systems in International Relations Research, pp. 199-215. Edited by John V. Gillespie and Dina A. Zinnes. New York: Praeger, 1977.

---. "International Power and the Distribution of World Wealth." In The New Economics of the L.D.C.: Changing Perceptions in the North-South Dialogue, pp. 303-318. Edited by Nake Kamrany. Boulder, CO: Westview Press, 1978.

---. "A Multiagent Adaptive Approach to Macroeconomic Control." Policy Analysis and Information Systems. 2:2 (1979): 1-9.

---. "A Game Theoretical Approach to Bureaucratic Behavior." In Dynamic Optimization and Mathematical Economics, pp. 223-236. Edited by P.T. Liu. New York: Plenum Press, 1980.

---. "SALT and Innovations in Weapons Technology." Public Choice. 36 (1981): 41-59.

---. "Arms Races: Behavioral and Economic Dimensions." In Missing Elements in Political Inquiry: Logics and Levels of Analysis, pp. 93-116. Edited by Judith A. Gillespie and Dina A. Zinnes. Beverly Hills, CA: Sage, 1982.

Brody, Richard A. "Some Systematic Effects of the Spread of Nuclear Weapons Technology: A Study Through Simulation of a Multi-nuclear Future." Journal of Conflict Resolution. 7:4 (1963): 663-753.

Brody, Richard A.; Benham, Alexandra H.; and Milstein, Jeffrey S. "Hostile International Communication, Arms Production, and Perception of Threat: A Simulation Study." Peace Research Society Papers. 7 (1968): 15-40.

Brooks, Harvey. "The Military Innovation System and the Qualitative Arms Race." Daedalus. 104:3 (Summer 1975): 75-98.

Brown, John Michael. "An Application of Richardson's Equations to U.S. and U.S.S.R. Military Outlays, 1948-1968." Ph.D. dissertation, University of Kentucky, 1971.

Brown, T.A. "Models of Strategic Stability." Santa Monica, CA: Rand Corp., Southern California Arms Control and Foreign Policy Seminar, 1971.

---. "What Is an Arms Race?" Santa Monica, CA: Rand Corp., Southern California Arms Control and Foreign Policy Seminars, 1973.

Brubaker, Earl R. "Economic Models of Arms Races: Some Reformulations and Extensions." Journal of Conflict Resolution. 17:2 (June 1973): 187-205.

Brzoska, Michael. "The Reporting of Military Expenditures." Journal of Peace Research. 18:3 (1981): 261-275.

Bueno de Mesquita, Bruce. "An Expected Utility Explanation of Conflict Escalation: A Preliminary Analysis." In Conflict Processes and the Breakdown of International Systems, pp. 47-59. Edited by Dina A. Zinnes. Denver: Graduate School of International Studies, 1983.

Bull, Hedley. The Control of the Arms Race. New York: Praeger, 1961, 1965.

Burns, Arthur Lee. "A Graphical Approach to Some Problems of the Arms Race." Journal of Conflict Resolution. 3:4 (December 1959): 326-342.

Busch, Peter. "Mathematical Models of Arms Races." In What Price Vigilance? The Burdens of National Defense, pp. 193-233. Edited by Bruce M. Russett. New Haven: Yale University Press, 1970.

Cady, R.H. "Some Notes on the Theory of Arms Races." (BSR 2063). Los Angeles: Bendix Corporation, 1966.

Carter, L.J. "Strategic Arms Limitation (II): 'Leveling VP' to Symmetry." Science. 187 (1975): 627-632.

Caspary, William R. "Richardson's Model of Arms Races: Description, Critique and an Alternative Model." International Studies Quarterly. 11:1 (March 1967): 63-88.

---. "Formal Theories of Reaction Processes." Paper presented to the annual meeting of the American Political Science Association, September 1969.

Chadwick, Richard W. "Modeling Political-Military Policy Dynamics in a Global Model." In Theories, Models, and Simulations in International Relations, pp. 115-132. Edited by Michael D. Ward. Boulder, CO: Westview Press, 1985.

Chase, P.E. "Control Theory and the Nuclear Arms Race." Bendix Technical Journal. (1968): 43-54.

---. "The Relevance of Arms Race Theory to Arms Control." General Systems. 13 (1968): 91-98.

---. "Feedback Control Theory and Arms Races." General Systems. 14 (1969): 137-149.

Chatterjee, Partha. "Balance of Power in Alliances and Arms Races." Ph.D. dissertation, University of Rochester, 1971.

---. "The Equilibrium Theory of Arms Races: Some Extensions." Journal of Peace Research. 11:3 (1974): 203-212.

---. Arms, Alliances and Stability: The Development of the Structure of International Politics. New York: John Wiley and Sons, 1975.

---. "The Equilibrium Theory of Arms Races: Reply to Comments by F.E. Banks." Journal of Peace Research. 12:3 (1975): 239-241.

Chatterji, Manas. "A Model of Resolution of Conflict between India and Pakistan." Peace Science Society (International) Papers. 12 (1969): 87-102.

Choucri, Nazli. "Defense Budgets and the Middle East Conflict: A Commentary." Peace Research Society (International) Papers. 25 (1970): 28-30.

---. "Applications of Econometric Analysis to Forecasting in International Relations." Peace Research Society (International) Papers. 21 (1973): 15-40.

Choucri, Nazli, and North, Robert C. "Aspects of International Conflict: Military Preparedness, Alliance Commitments, and External Violence." Stanford, CA: Studies in International Conflict and Integration, April 1969.

---. "The Determinants of International Violence." Peace Research Society Papers. 12 (1969): 33-63.

---. Nations in Conflict: National Growth and International Violence. San Francisco: W.H. Freeman and Co., 1975.

Cioffi-Revilla, Claudio A. "Theories of Arms Races and Mathematical Structures in International Relations." Paper presented to the joint meeting of the Mid-west Political Science Assocation and the Mid-west section of the Peace Science Society (International), Chicago, April 19-20, 1979.

---. "Political Reliability in International Relations." In Conflict Processes and the Breakdown of International Systems, pp. 11-45. Edited by Dina A. Zinnes. Denver: Graduate School of International Studies, 1983.

Cioffi-Revilla, Claudio; Muncaster, Robert; and Zinnes, Dina A. "Defining War." Paper presented to the annual meeting of the International Studies Association, Atlanta, March 27-31, 1984.

Clarke, T. Michael. "Modeling External and Internal Conflict in Small Countries." Paper presented to the Conference on Dynamic Models of Conflict, Boulder, CO, October 31 to November 3, 1984.

Clarke, T. Michael; Luterbacher, Urs; and Allan, Pierre. "International Public Policy Analysis: Energy Crises and the Small State." In Theories, Models, and Simulations in International Relations, pp. 79-100. Edited by Michael D. Ward. Boulder, CO: Westview Press, 1985.

Cory, Robert H., Jr. "Referees in the Armaments Race: International Inspection Teams." Journal of Conflict Resolution. 3:4 (December 1959): 417-419.

Crecine, John P. "Defense Budgeting: Organizational Adaptation to Environmental Constraints." In Studies of Budgeting, pp. 210-261. Edited by R.F. Byrnes et al. Amsterdam: North-Holland Publishing Co., 1971.

Crecine, John P., and Fischer, Gregory W. "On Resource Allocation in the U.S. Department of Defense." In Political Science Annual: An International Review, Vol. 4, pp. 181-236. Edited by C.P. Potter. Indianapolis: Bobbs-Merrill, 1973.

Crookall, David. "The ICONS Foreign Policy/Foreign Language Simulation." Paper presented to the annual meeting of the International Studies Association, Anaheim, CA, March 25-29, 1986.

Crow, Wayman J. "A Study of Strategic Doctrines Using the Inter-nation Simulation." Journal of Conflict Resolution. 7:3 (September 1963): 580-589.

Cusack, Thomas R. "The Evolution of Power, Threat and Security: Past and Potential Developments." Paper presented to the annual meeting of the International Studies Association, Atlanta, March 27-31, 1984.

---. "One Problem, Three Solutions: A Simulation Analysis of Alternative Western Defense Policy Options." IIVG Discussion Paper 84-109. Berlin: Wissenscaftszentrum Berlin, November 1984.

Cusack, Thomas R., and Stoll, Richard J. "Rationality, Aggression, and Endurance in an Anarchic World: A Computer Simulation." Paper presented to the annual meeting of the International Studies Association, Washington, DC, April 14-18, 1987.

Cusack, Thomas R., and Ward, Michael Don. "Military Spending in the United States, Soviet Union, and the People's Republic of China." Journal of Conflict Resolution. 25:3 (September 1981): 429-469.

Dacey, Michael F. "A Probability Model for the Rise and Decline of States." Peace Research Society Papers. 14 (1970): 147-153.

Dacey, Raymond. "Detection and Disarmament: A Comment on the Geometry of the Arms Race." International Studies Quarterly. 23:4 (December 1979): 589-598.

Dacey, Raymond, and Pendegraft, Norman. "The Impact of Election Frequency and Crises upon the Arms Race." Paper presented to the annual meeting of the International Studies Association, Washington, DC, April 14-18, 1987.

Dalkey, Norman. "Solvable Nuclear War Models." RM-4009-PR. Santa Monica, CA: Rand Corp., April 1964.

Dash, J.G. "A Condensation in Warpeace Space." General Systems. 11 (1966): 143-146.

---. "Comments on the Paper by Smoker." Peace Research Society Papers. 7 (1967): 63-65.

Davis, R.H.; Carpenter, P.B.; and Missler, C.W. "A Game for Studying the Problem of Arms Control." General Systems. 8 (1963): 83-89.

Davis, William W.; Duncan, George T.; and Siverson, Randolph M. "Stochastic Models of the Distribution of Dyadic War in Time." Technical Report No. 120. Pittsburgh: Department of Statistics, Carnegie-Mellon University, 1976.

---. "The Dynamics of Warfare: 1816-1965." American Journal of Political Science. 22:4 (November 1978): 772-792.

Dennis, Michael F.H. "The Policy Basis of General Purpose Forces." Journal of Conflict Resolution. 18:1 (March 1974): 3-36.

Denton, Frank Hardy. "Patterns of Political Violence and War, 1751-1960." Ph.D. dissertation, University of Southern California, 1968.

Despres, J., and Dhyrmes, P. "Defense Budget Interactions: Preliminary Econometric Analysis of U.S., Soviet and Chinese Competition." Confidential, R-1390-PR. Santa Monica, CA: Rand Corp., 1973.

Diehl, Paul F. "Arms and Escalation: A Closer Look." Journal of Peace Research. 20:3 (1983): 205-212.

---. "Arms Races to War: Testing Some Empirical Linkages." Paper presented to the annual meeting of the American Political Science Association, Chicago, September 1-4, 1983.

Dolansky, L. "Present State of the Lanchester Theory of Combat." Operations Research. 12:2 (1964): 344-358.

Doran, C. "A Theory of Bounded Deterrence." Journal of Conflict Resolution. 17 (1973): 243-269.

Doty, Paul et al. "The Race to Control Nuclear Arms." Foreign Affairs. 55 (October 1976): 119-132.

Dumas, Lloyd J. "Comments on 'A Catastrophe Surface in Mutual Deterrence Theory.'" Journal of Peace Science. 3:2 (Fall 1978): 123-128.

Duncan, George T., and Siverson, Randolph M. "Markov Chain Models for Conflict Analysis: Results from Sino-Indian Relations, 1959-1964." International Studies Quarterly. 19:3 (September 1975): 344-374.

Dupuy, T.N. Numbers, Prediction and War: Using History to Evaluate Combat Factors and Predict the Outcome of Battles. London: MacDonald and Jane's, 1979.

Eberwein, Wolf-Dieter. "The Quantitative Study of International Conflict: Quantity or Quality." Journal of Peace Research. 18:1 (1981): 19-38.

Eckhardt, William. "Peace Research Agenda for the 1980's." International Studies Newsletter. 11:5 (June/July 1984): 14-15.

Etcheson, Craig. "Strategy and the Structure of Behavior in Arms Races." Ph.D. dissertation, University of Southern California, 1985.

---. "Strategy in the Arms Race: A Qualitative Computer Simulation." Paper presented to the annual meeting of the International Studies Association, Washington, DC, March 5-9, 1985.

---. "Strategy and Arms Accumulation Research." Paper presented to the annual meeting of the American Political Science Association, Chicago, September 3-6, 1987.

Etcheson, Craig, and Scacchi, Walt. "MONADYSIM: Applying AI Techniques to the Study of War." Proceedings of the IASTED International Symposium: Applied Simulation and Modeling, pp. 35-38. Anaheim, CA: ACTA Press, 1982.

Evangelista, Matthew A. "Technological Innovation and the Arms Race: A Comparative Study of Soviet and American Decisions on Tactical Nuclear Weapons." Ph.D. dissertation, Cornell University, 1986.

Fedder, E.H. "On the Use of Models in Simulations and Games." International Studies Quarterly. 13 (1969): 111-116.

Fei, E.T. "Understanding Arms Transfers and Military Expenditures: Data Problems." In Arms Transfers in the Modern World, pp. 37-48. Edited by S.G. Neuman and R.E. Harkavy. New York: Praeger, 1979.

Feigl, Hubert. "Die Technischen Komponenten der Rustungsdynamik: Technologische Aspekte der Gesprache Uber Beschrankungen Straegische Waffen." Europa Archiv. 24:21 (November 10, 1969): 757-768.

Feld, B.T. et al., eds. Impact of Technologies on the Arms Race. Cambridge, MA: MIT Press, 1971.

Ferejohn, John A. "On the Effects of Aid to Nations in Arms Races." In Mathematical Models in International Relations, pp. 218-251. Edited by Dina A. Zinnes and John V. Gillespie. New York: Praeger, 1976.

Flora, Peter. "A New Stage of Political Arithmetic." Journal of Conflict Resolution. 18:1 (1974): 143-165.

Freeman, John. "Granger Causality and the Time Series Analysis of Political Relationships." American Journal of Political Science. 27:2 (May 1983): 327-358.

Friberg, Mats, and Jonsson, Dan. "A Simple War and Armaments Game." Journal of Peace Research. 5 (July 1968): 233-247.

Gallois, P. The Balance of Terror. Boston: Houghton-Mifflin, 1961.

Gantzel, Klause Jurgen. "Armament Dynamics in the East-West Conflict: An Arms Race?" Peace Science Society (International) Papers. 20 (1973): 1-24.

---. "Another Approach to a Theory on the Causes of International War." Journal of Peace Research. 18:1 (1981): 39-55.

Gerner, Debbie J. "The Effects of International Arms Transfers on Domestic Conflict in Africa and the Middle East." Paper presented to the annual meeting of the International Studies Association, Mexico City, April 5-9, 1983.

Gillespie, John V. "Optimal Control Theory: A Promising Approach for Future Research." In In Search of Global Patterns, pp. 235-246. Edited by James N. Rosenau. New York: Free Press, 1976.

---. "Why Mathematical Models?" In Mathematical Models in International Relations, pp. 37-61. Edited by Dina A. Zinnes and John V. Gillespie. New York: Praeger, 1976.

Gillespie, John V., and Rubison, R. "An Optimal Control Model of Arms Races and Its Implications for SALT." Paper presented to the annual meeting of the Mid-west Political Science Association, Chicago, April 26, 1974.

Gillespie, John V., and Zinnes, Dina A. "Progressions in Mathematical Models of International Conflict." Synthese. 31 (1975): 289-321.

---, eds. Mathematical Systems in International Relations Research. New York: Praeger, 1977.

Gillespie, John V.; Zinnes, Dina A.; and Rubison, R. Michael. "Accumulation in Arms Race Models: A Geometric Lag Perspective." Comparative Political Studies. 10:4 (January 1978): 475-496.

Gillespie, John V.; Zinnes, Dina A.; Schrodt, Philip A.; and Tahim, G.S. "A Sensitivity Analysis of a Perceptual Arms Race Model." Paper presented to the annual meeting of the Mid-west Political Science Association, Chicago, April 26, 1974.

---. "Sensitivity Analysis of an Armaments Race Model." In Threats, Weapons and Foreign Policy, pp. 275-310. Edited by Pat McGowan and Charles W. Kegley, Jr. Beverly Hills, CA: Sage, 1980.

Gillespie, John V.; Zinnes, Dina A.; and Tahim, G.S. "Foreign Military Assistance and the Armaments Race: A Differential Game Model with Control." Peace Science Society (International) Papers. 25 (1975): 35-51.

---. "Deterrence as Second Attack Capability: An Optimal Control Model and Differential Game." In Mathematical Systems in International Relations Research, pp. 367-385. Edited by John V. Gillespie and Dina A. Zinnes. New York: Praeger, 1977.

Gillespie, John V.; Zinnes, Dina A.; Tahim, G.S.; Sampson, Martin W. III; Schrodt, Philip A.; and Rubison, R.M. "Deterrence and Arms Races: An Optimal Control Systems Model." Behavioral Science. 24:4 (July 1979): 250-262.

Gillespie, John V.; Zinnes, Dina A.; Tahim, G.S.; Schrodt, Philip A.; and Rubison, R.M. "An Optimal Control Model of Arms Races." American Political Science Review. 71:1 (1977): 226-244.

Gleditsch, N.P. "Fragments of a Theory of Arms Races." (BSR 2059). Los Angeles: Bendix Corporation, 1967.

Gold, E. "Lewis Fry Richardson." Obituary Notices of Fellows of the Royal Society. 9 (1954): 217-235.

Goldblat, Josef, and Millan, Victor. The Falklands/Malvinas Conflict: A Spur to Arms Build-ups. Stockholm: SIPRI, 1983.

Gray, Colin S. "Action and Reaction in the Nuclear Arms Race." Military Review. 51 (August 1971): 16-26.

---. "The Arms Race Phenomenon." World Politics. 24:1 (October 1971): 39-79.

---. "'Gap' Prediction and America's Defense Arms Race Behavior in the Eisenhower Years." Orbis. 16:1 (Spring 1972): 257-274.

---. "Social Science and the Arms Race." Bulletin of the Atomic Scientists. 29:6 (June 1973): 23-26.

---. "The Arms Race is About Politics." Foreign Policy. 9 (Winter 1973): 117-129.

---. "The Urge to Compete: Rationales for Arms Racing." World Politics. 26:2 (January 1974): 207-233.

---. "Arms Races and Their Influence upon International Stability with Special Reference to the Middle East." In Dynamics of Conflict. Edited by G. Sheffer. New York: Rand McNally, 1975.

---. The Soviet-American Arms Race. Lexington, MA: Lexington Books, 1976.

---. The Future of Land-Based Missile Forces. Adelphi Paper No. 140. London: International Institute for Strategic Studies, 1977.

Green, Phillip. Deadly Logic. Columbus: Ohio State University Press, 1966.

Gregory, P. "Economic Growth, U.S. Defense Expenditures, and the Soviet Budget." Soviet Studies. 24 (1974): 72-80.

Griffin, L.J.; Wallace, Michael; and Devine, J. "The Political Economy of Military Spending and Evidence from the United States." 6 (1982): 1-14.

Groover, P.L.; Garrison, W.V.; and Cotsworth, W.L. The Arsenal Exchange Model. Denver: Martin-Marietta, 1971.

Guetzkow, Harold; Hollist, W. Ladd; and Ward, Michael Don. "Computer Simulations for Decision-Making in International Affairs: Moving Toward Consolidation of Research in International Relations Through Empirical Analysis and Computer Simulation." International Peace Research Newsletter. 15:3 (1977): 5-12.

Guetzkow, Harold, and Valadez, Joseph J. Simulated International Processes: Theories and Research in Global Modeling. Beverly Hills, CA: Sage, 1981.

Hamblin, R.L.; Hout, M.; Miller, J.L.L.; and Pitcher, B.L. "Arms Races: A Test of Two Models." American Sociological Review. 42 (April 1977): 338-354.

Harsanyi, John. "Mathematical Models for the Genesis of War." World Politics. 14 (1962): 687-699.

---. "Game Theory and the Analysis of International Conflict." In International Politics and Foreign Policy, pp. 370-379. Edited by James N. Rosenau. New York: Free Press, 1969.

Hausrath, Alfred H. Venture Simulation in War, Business, and Politics. New York: McGraw-Hill, 1971.

Hermann, Charles F. "Validation Problems in Games and Simulations with Special Reference to Models of International Politics." Behavioral Science. 12:3 (May 1967): 216-231.

Hermann, Charles F., and Hermann, Margaret G. "The Synthetic Role of Decision-Making Models in Theories of Foreign Policy: Bases for a Computer Simulation." In Theories, Models, and Simulations in International Relations, pp. 223-248. Edited by Michael D. Ward. Boulder, CO: Westview Press, 1985.

Hill, Walter. "A Time Lagged Arms Race Model." Journal of Peace Science. 3:1 (Spring 1978): 55-62.

Hoag, Malcolm W. "On Stability in Deterrent Races." World Politics. 13:4 (July 1961): 505-527.

Hoffmann, F. "Arms Debates: A Positional Interpretation." Journal of Peace Research. 3 (1970): 219-228.

Hollist, W. Ladd. "Alternative Explanation of Competitive Arms Processes: Tests on Four Pairs of Nations." American Journal of Political Science. 21:2 (May 1977): 313-340.

---. "An Analysis of Arms Processes in the United States and the Soviet Union." International Studies Quarterly. 21:3 (September 1977): 503-528.

---, ed. Exploring Competitive Arms Processes: Applications of Mathematical Modeling and Computer Simulation in Arms Policy Analysis. New York: Marcel Dekker, 1978.

Hollist, W. Ladd., and Guetzkow, Harold. "Cumulative Research in International Relations: Empirical Analysis and Computer Simulation of Competitive Arms Processes." In Exploring Competitive Arms Processes, pp. 165-197. Edited by W. Ladd Hollist. New York: Marcel Dekker, 1978.

Hollist, W. Ladd, and Johnson, Thomas H. "Political-Economic Competition between the United States and the Soviet Union: Three Alternative Simulations." Paper presented at the western meeting of the Peace Science Society (International), Stanford, CA, February 26-27, 1979.

Hollist, W. Ladd; Johnson, Thomas H.; and Boydston, Robert J. "Exploring US-USSR Arms Race Assumptions: What-If Queries via Computer Simulation." Paper presented to the annual meeting of the International Studies Association, Washington, DC, February 22-25, 1978.

Holst, Johan Jorgen. "What Is Really Going On?" Foreign Affairs. 19 (Summer 1975): 155-163.

Holsti, Ole; Brody, Richard A.; and North, Robert C. "Measuring Affect and Action in International Reaction Models: Empirical Material from the 1962 Cuban Missile Crisis." Journal of Peace Research. 3 (1964): 170-190.

Holt, Robert T.; Job, Brian L.; and Markus, Lawrence. "Catastrophe Theory and the Study of War." Journal of Conflict Resolution. 22:2 (June 1978): 171-208.

Hoole, Frank W.; Ostrom, Charles W., Jr.; and Tucker, Harvey J. "Toward a More Complete Model of Budgeting in International Organization." Paper presented to the annual meeting of the International Studies Association, Los Angeles, March 19-22, 1980.

---. "Budgetary Model Complexity and International Organizations." In Theories, Models, and Simulations in International Relations, pp. 305-328. Edited by Michael D. Ward. Boulder, CO: Westview Press, 1985.

Hopmann, Terry, and Smith, T. "An Application of a Richardson Process Model: Soviet-American Interactions in the Test Ban Negotiations 1962-63." Journal of Conflict Resolution. 21:4 (December 1977): 701-726.

Horn, Mike. "Arms Races and the Likelihood of War." Paper presented to the annual meeting of the International Studies Association, Atlanta, March 27-31, 1984.

Horvath, William J. "A Statistical Model for the Duration of Wars and Strikes." Behavioral Science. 13:1 (January 1968): 18-28.

Horvath, William J., and Foster, Caxton C. "Stochastic Models of War Alliances." Journal of Conflict Resolution. 7:1 (1963): 110-116.

Huckfeldt, R. Robert; Kohfeld, C.W.; and Likens, Thomas W. Dynamic Modeling: An Introduction. Beverly Hills, CA: Sage, 1982.

Hughes, Barry B. "The International Futures Simulation." Paper presented to the annual meeting of the International Studies Association, Anaheim, CA, March 25-29, 1986.

Hunter, D. "Decision-Making Model in Nuclear Deterrence Theory." Journal of Peace Research. 3 (1972): 209-222.

Hunter, John E. "Mathematical Models of a Three-Nation Arms Race." Journal of Conflict Resolution. 24:2 (June 1980): 241-252.

Huntington, Samuel P. "Arms Races: Prerequisites and Results." Public Policy. 8 (1958): 41-86.

Hutchings, Raymond. "Soviet Defense Spending and Soviet External Relations." International Affairs (London). 47 (1971): 518-531.

Hybel, Alex R., and Robertson, Dale B. "Assessing the Dynamic Reaction of the USSR to American and Chinese Actions and to Its Nuclear Gap with the USA." International Interactions. 4:2 (1978): 125-154.

Intriligator, Michael D. "Some Simple Models of Arms Races." General Systems. 9 (1964): 143-147.

---. Strategy in a Missile War: Targets and Rates of Fire. Paper No. 10. Los Angeles: UCLA Security Studies Project, 1967.

---. "Arms Races and War Initiation: The Effect of Strategic Choices." Report # WEC-126:6:31. Washington, DC: U.S. Arms Control and Disarmament Agency, 1968.

---. "The Debate Over Missile Strategy: Targets and Rates of Fire." Orbis. 11:4 (Winter 1968): 1138-1159.

---. Mathematical Optimization and Economic Theory. Englewood Cliffs, NJ: Prentice-Hall, 1971.

---. "Applications of Optimal Control Theory in Economics." Synthese. 31 (1975): 271-288.

---. "Strategic Considerations in the Richardson Model of Arms Races." Journal of Political Economy. 83:2 (April 1975): 339-353.

---. Econometric Models, Techniques, and Applications. Englewood Cliffs, NJ: Prentice-Hall, 1978.

---. "Research on Conflict Theory: Analytical Approaches and Areas of Application." Journal of Conflict Resolution. 26:2 (1982): 307-327.

Intriligator, Michael D., and Brito, Dagobert L. "Formal Models of Arms Races." Journal of Peace Science. 2:1 (1976): 77-88.

---. "Strategy, Arms Races, and Arms Control." In Mathematical Systems in International Relations Research, pp. 173-189. Edited by John V. Gillespie and Dina A. Zinnes. New York: Praeger, 1977.

---. "Nuclear Proliferation and Stability." Journal of Peace Science. 3:2 (1978): 173-183.

---. "Nuclear Proliferation and the Probability of Nuclear War." Public Choice. 37 (1981): 247-260.

---. "Can Arms Races Lead to the Outbreak of War?" Journal of Conflict Resolution. 28:1 (1984): 63-84.

---. "Heuristic Decision Rules, the Dynamics of the Arms Race, and War Initiation." Paper presented to the conference on Dynamic Models of Conflict, Boulder, CO, October 31 to November 3, 1984.

---. "Non-Armageddon Solutions to the Arms Race." CISA Reprint No. 1. Los Angeles: University of California, Center for International Strategic Affairs, 1984.

---. "Arms Control, Disarmament, and Global Peace and Security." Paper presented to the annual meeting of the International Studies Association, Washington, DC, March 5-9, 1985.

---. "The Stability of Mutual Deterrence." N.p.: (Mimeo), 1985.

Isard, Walter. "Social Systems Framework and Causal History: Part I, Learning Processes." Conflict Management and Peace Science. 6:1 (Fall 1981): 63-94.

---. "Social Systems Framework and Causal History: Part II, Behavior Under Stress and Cognition." Conflict Management and Peace Science. 6:2 (Spring 1982-83): 59-93.

---. "Social Systems Framework and Causal History: Part III, Interdependent Multi-role Behavior and Operationality." Conflict Management and Peace Science. 7:1 (Fall 1983): 87-110.

---. "On the Incorporation of Arms Race Models in World Systems Models." Paper presented to the annual meeting of the International Studies Association, Washington, DC, March 5-9, 1985.

Isard, Walter, and Smith, T.E. "Practical Applications of Game Theoretical Approaches to Arms Reduction." Peace Research Society Papers. 4 (1965): 128-132.

Ivanilov, Yu P. "A Model of Competition between Two Countries." Engineering Cybernetics. 1 (January-February 1969): 19-30.

James, Patrick. "Crisis, War, and Expected Utility: Why Do Some Crises Escalate to War While Others Do Not." Paper presented to the annual meeting of the International Studies Association, Anaheim, CA, March 25-29, 1986.

Job, Brian L. "Membership in Inter-nation Alliances, 1815-1965: An Exploration Utilizing Mathematical Probability Models." In Mathematical Models in International Relations, pp. 74-109. Edited by Dina A. Zinnes and John V. Gillespie. New York: Praeger, 1976.

---. "Dynamic Modeling: Introduction and Synthesis." In Missing Elements in Political Inquiry, pp. 89-9, 141-149. Edited by Judith A. Gillespie and Dina A. Zinnes. Beverly Hills, CA: Sage, 1982.

Joenniemi, Pertti. "Aspects on the Measurement of Armaments Levels." Cooperation and Conflict. 5:3 (1970): 141-151.

Joynt, C.B. "Arms Races and the Problem of Equilibrium." Yearbook of World Affairs. 18 (1964): 23-40.

Kahn, Herman. "The Arms Race and Some of Its Hazards." Daedalus. 89 (Fall 1960): 744-780.

---. "The Arms Race and World Order." In The Revolution in World Politics, pp. 332-351. Edited by Morton Kaplan. New York: Wiley, 1962.

Kalman, Peter J. "Application of a Theorem by Scarf on Cooperative Solutions to N-Person Games to Problems of Disarmament." Journal of Political Economy. 79:6 (November/December 1971): 1388-1396.

Kaplowitz, Noel. "Rational Actor Models and Social Psychological Models in the Interpretation of War and Peace." Paper presented to the annual meeting of the International Studies Association, Anaheim, CA, March 25-29, 1986.

Karmeshu, K. "Statistical Study of the Richardson's Arms Race Model with Time Lag." Conflict Management and Peace Science. 4 (1980): 69-78.

Kemp, Anita. "A Computer Model of International Violence." International Interactions. 3:1 (1977): 51-62.

---. "A Path Analytic Model of International Violence." International Interactions. 4:1 (1977): 53-85.

Kenda, I. "Dynamics of War, Arms Trade, and Military Expenditure." Helsinki: Symposium on Armament, Tension and War, September 1977.

Kent, Glenn A. On the Interaction of Opposing Forces Under Possible Arms Agreements. Cambridge, MA: Harvard University, Center for International Affairs, 1963.

Kirkpatrick, Grant H., and Widmaier, Ulrich. "Linking Islands of Theory and Technique in Political Economy." In Theories, Models, and Simulations in International Relations, pp. 133-179. Edited by Michael D. Ward. Boulder, CO: Westview Press, 1985.

Kobe, Donald H. "A Theory of Catalytic War." Journal of Conflict Resolution. 6:2 (June 1962): 125-142.

Kohfeld, C.W., and Salert, B. "Discrete and Continuous Representation of Dynamic Models." Political Methodology. 8 (1982): 1-32.

Kohler, Gernot. "Une theorie structuro-dynamique des armements." Etudies internationales (Canada). 7:1 (March 1976): 25-50.

---. "Exponential Military Growth." Peace Research (Canada). 9:4 (1977): 165-175.

---. "Toward a General Theory of Armaments." Journal of Peace Research. 16:2 (1979): 117-135.

Kugler, J. "Models of Deterrence." Paper presented to the annual meeting of the Mid-west Political Science Association, Chicago, 1979.

Kugler, J., and Organski, A.F.K., with Fox, D.J. "Deterrence and the Arms Race: The Impotence of Power." International Security. 1 (1980): 105-138.

Kupperman, Robert H., and Smith, Harvey A. "Some Computational Techniques and Theorems Concerning Mutual Deterrence Relationships." Journal of Cybernetics. 1:3 (1971): 5-28.

---. "Strategies of Mutual Deterrence." Science. 176 (April 7, 1972): 18-23.

---. "Formal Models of Arms Races: Discussion." Journal of Peace Science. 2:1 (1976): 89-96.

---. "Deterrent Stability and Strategic Warfare." In Mathematical Systems in International Relations Research, pp. 139-166. Edited by John V. Gillespie and Dina A. Zinnes. New York: Praeger, 1977.

---. "The Role of Population Defense in Mutual Deterrence." SIAM Review. 19:2 (April 1977): 297-318.

---. "A Catastrophe Surface in Mutual Deterrence Theory: Arms Limitation and Crisis Instability." Journal of Peace Science. 3:2 (Fall 1978): 111-121.

Kupperman, Robert H.; Smith, Harvey A.; and Abramson, L.R. "On Strategic Intelligence in Deterrence." Journal of Cybernetics. 2 (1972): 3-20.

Kupperman, Robert H.; Smith, Harvey A.; Goldman, Steven C.; and Johnson, Mark R. "Some Control Theoretic Prospects for International Relations." In Mathematical Systems in International Relations Research, pp. 36-48. Edited by John V. Gillespie and Dina A. Zinnes. New York: Praeger, 1977.

Kurokawa, S. "A Simple Model of Arms Races: Richardson's Model Revisited." Paper presented to the Japan conference of the Peace Science Society (International), Hiroshima, 1978.

Lagerstrom, Richard P., and North, Robert C. "An Anticipated Gap, Mathematical Model of International Dynamics." Stanford, CA: Studies in International Conflict and Integration, February 8, 1968.

Lambelet, Jean-Christian. "A Dynamic Model of the Arms Race in the Middle East, 1953-1965." General Systems. 16 (1971): 145-167.

---. "Towards a Dynamic Manpower-and-Hardware Model of the East-West Arms Race: First Results." Discussion Paper No. 241. Philadelphia: University of Pennsylvannia, Department of Economics, 1972.

---. "Towards a Dynamic Two-Theater Model of East-West Arms Race." Journal of Peace Science. 1 (1973): 1-37.

---. "The Anglo-German Dreadnought Race, 1905-1914." Peace Science Society (International) Papers. 22 (1974): 1-45.

---. "Do Arms Races Lead to War?" Journal of Peace Research. 12:2 (1975): 123-128.

---. "A Numerical Model of the Anglo-German Dreadnought Race." Peace Science Society (International) Papers. 24 (1975): 29-48.

---. "A Complementary Analysis of the Anglo-German Dreadnought Race, 1905-1916." Peace Science Society (International) Papers. 26 (1976): 49-66.

---. "Arms Races as Good Things?" Paper presented to the Conference on Dynamic Models of Conflict, Boulder, CO, October 31 to November, 3, 1984.

Lambelet, Jean-Christian, and Luterbacher, Urs, with Allan, Pierre. "Dynamics of Arms Races: Mutual Stimulation vs. Self Stimulation." Journal of Peace Science. 4 (1979): 49-66.

Lee, John Ryool, and Milstein, Jeffrey S. "A Political Economy of the Vietnam War, 1965-1972." Peace Science Society (International) Papers. 21 (1973): 41-63.

Legault, Albert, and Lindsey, George. Dynamik des Nuklearen Gleichgewichtes. Frankfurt: Metzen, 1973.

Leitenberg, Milton. "The Present State of the World Arms Race." Cooperation and Conflict. 5:3 (1970): 185-197.

Levy, Marion J., Jr. "A Review of Anatol Rapoport's Fights, Games and Debates." Journal of Conflict Resolution. 6:1 (March 1962): 96-99.

Li, Richard Y.P., and Thompson, William R. "The 'Coup Contagion' Hypothesis." Journal of Conflict Resolution. 19:1 (1975): 63-88.

Liossatos, P. "Modeling the Nuclear Arms Race: A Search for Stability." Journal of Peace Science. 4 (1980): 169-185.

Lodgaard, Steve. "Arms Race in Europe." Journal of Peace Research. 17:1 (1980): 1-8.

Lucier, Charles. "Changes in the Values of Arms Race Parameters." Journal of Conflict Resolution. 23:1 (March 1979): 17-39.

Luterbacher, Urs. Dimensions Historiques de Modeles Dynamiques de Conflit: Application aux processus de course aux armements 1900-1965. Leiden: Sijthoff, 1974.

---. "Arms Race Models: Where Do We Stand?" European Journal of Political Research. 3 (1975): 199-217.

---. "Towards a Convergence of Behavioral and Strategic Conceptions of the Arms Race: The Case of American and Soviet ICBM Build-Up." Peace Science Society (International) Papers. 26 (1976): 1-21.

---. "Simpest: A Simulation Model of Interactions among Major Powers." Geneva: C.R.E.R.I., 1980.

---. "Global Modeling, The Difficult Art of Trying to Grasp World Problems: A Review and Suggestions for Further Research." Geneva: (Mimeo), pp. 207-224.

Luterbacher, Urs, and Allan, Pierre. "Modeling Politico-Economic Interactions within and between Nations." International Political Science Review. 3:4 (1982): 404-433.

---. "Les Consequences du Programme de Defense du President Reagan." Geneva: C.R.E.R.I., 1983.

---. "Shifting Alignments in the Triad: The Case of the US, the USSR, and China." Paper presented to the annual meeting of the International Studies Association, Mexico City, April 5-9, 1983.

---. "Simulating the Initiation, Escalation, and Termination of Conflict." Paper presented to the UCLA Summer Institute for the Study of Conflict Theory and International Security, Los Angeles, 1983.

---. "Towards a Dynamic Theory of Conflict and Coalition Formation." Paper presented to the Conference on Dynamic Models of Conflict, Boulder, CO, October 29 to November 3, 1984.

Luterbacher, Urs; Allan, Pierre; and Imhoff, Allen. "SIMPEST: A Simulation Model of the Military, Political, Economic and Strategic Interactions of the Major Powers." Paper presented to the world congress of the International Political Science Association, Moscow, August 12-18, 1979.

Luterbacher, Urs; Clarke, T. Michael; and Allan, Pierre. "SIMSWISS: A Simulation Model of Switzerland for Planning." Report to the Swiss National Science Foundation, Geneva, n.d. 1983.

Luterbacher, Urs, and Imhoff, Andre. "Modeling the Politico-Economic System: The Case of the U.S." Geneva: C.R.E.R.I., n.d.

Luterbacher, Urs, and Lambelet, Jean-Christian. "Dynamics of Arms Races: Mutual Stimulation vs. Self-Stimulation." Paper presented to the annual European meeting of the Peace Science Society (International), Geneva, September 4-9, 1978.

Luterbacher, Urs, and Ward, Michael D., eds. Dynamic Models of International Conflict. Boulder, CO: Lynne Rienner, 1985.

McCormick, James N. "Evaluating Models of Crisis Behavior." International Studies Quarterly. 19:1 (March 1975): 17-45.

McCubbins, Mathew D. "A Decision-Theoretic Approach to Arms Competition." Social Science Working Paper No. 194. Pasadena: California Institute of Technology, November 1979.

---. "Economic Models of Arms Control." Austin: University of Texas, November 1982, (Mimeo).

---. "Policy Components of Arms Competition." American Journal of Political Science. 27:3 (August 1983): 385-406.

McCubbins, Mathew, and Williamson, P. "Strategic Arms Race Interaction and the Richardson Model: Tentative Results." Paper presented to the annual Western meeting of the Peace Science Society (International), Stanford, CA, February 27-28, 1978.

McGinnis, Michael. "Arms, Aid, and Allies--Technical Summary: The Regional Security Game Model." Minneapolis: University of Minnesota, n.d. [1984].

McGuire, Martin C. Secrecy and the Arms Race: A Theory of the Accumulation of Strategic Weapons and How Secrecy Affects It. Cambridge, MA: Harvard University Press, 1965.

---. "A Quantitative Study of the Strategic Arms Race in the Missile Age." Review of Economics and Statisitics. 59:3 (August 1977): 328-339.

---. "U.S. Assistance, Israeli Allocation, and the Arms Race in the Middle East: An Analysis of Three Interdependent Resource Allocation Processes." Journal of Conflict Resolution. 26 (1982): 199-235.

---. "Theoretical Notes on Lanchester Models of Military Tactics, Troop Density, Organization, and Attrition." Paper presented to the Conference on Dynamic Models of Conflict, Boulder, CO, October 31 to November 3, 1984.

Macrae, John, and Smoker, Paul. "A Vietnam Simulation: A Report on the Canadian/English Joint Project." Journal of Peace Research. 2:1 (1967): 1-25.

Majeski, Stephen J. "Mathematical Models of the U.S. Military Expenditure Decision-Making Process." Ph.D. dissertation, Indiana University, 1981.

---. "The Role of Anticipation and Expectation in Arms Race Processes." Paper presented to the annual meeting of the International Studies Association, Mexico City, April 5-9, 1983.

---. "Mathematical Models of the U.S. Military Expenditure Decision-Making Process. American Journal of Political Science. 27:3 (August 1983): 485-514.

---. "Dynamic Properties of the U.S. Military Expenditure Decision-Making Process." Conflict Management and Peace Science. 7:1 (Fall 1983): 65-86.

---. "Arms Races as Iterated Prisoner's Dilemma Games." Mathematical Social Sciences. 7 (1984): 253-266.

---. "Expectations and Arms Races." Paper presented to the annual meeting of the International Studies Association, Atlanta, March 27-31, 1984.

---. "Technological Innovation and Cooperation in Arms Races." International Studies Quarterly. 30:2 (June 1986): 175-191.

---. "Intervention and Neutralization: A Simulation Model of the 1961 Decisions in Vietnam and Laos." Paper presented to the annual meeting of the International Studies Association, Washington, DC, April 14-18, 1987.

Majeski, Stephen J., and Jones, David L. "Arms Race Modeling: Causality Analysis and Model Specification." Paper presented to the annual meeting of the International Studies Association, Los Angeles, March 19-22, 1980.

---. "Arms Race Modeling: Causality Analysis and Model Specification." Journal of Conflict Resolution. 25:2 (1981): 259-288.

Mandel, Robert. "Policy-Making Perspectives on War Simulations." Journal of Conflict Resolution. 24:2 (June 1980): 359-375.

---. "Professional-Level War Gaming: A Critical Assessment." In Theories, Models, and Simulations in International Relations, pp. 483-500. Edited by Michael D. Ward. Boulder, CO: Westview Press, 1985.

Marra, Robin F. "A Cybernetic Model of the US Defense Expenditure Policymaking Process." International Studies Quarterly. 29:4 (December 1985): 357-384.

Mayer, Thomas F. "Transform Methods and Dynamic Models." Paper presented to the Conference on Dynamic Models of Conflict, Boulder, CO, October 31 to November 3, 1984.

Meneses, Emilio. "Arms Competition in South America: A New Concept on Its Evaluation." Paper presented to the annual meeting of the International Studies Association, Mexico City, April 5-9, 1983.

Midgaard, Knut. "Arms Races, Arms Control, and Disarmament." Cooperation and Conflict. 5:1 (February 1970): 20-25.

Midlarsky, Manus I. "Mathematical Models of Instability and a Theory of Diffusion." International Studies Quarterly. 14:1 (March 1970): 60-84.

---. "Power and Distance in International Conflict Behavior." In Mathematical Models of International Relations, pp. 132-155. Edited by Dina A. Zinnes and John V. Gillespie. New York: Praeger, 1976.

---. "Alliance Behavior and the Approach to World War I: The Use of Bivariate Negative Binomial Distributions." In Conflict Processes and the Breakdown of International Systems, pp. 61-80. Edited by Dina A. Zinnes. Denver: Graduate School of International Studies, 1983.

---. "Predicting Systemic War: Patterns of Conflict Behavior Preceding the Onset of General War." Paper presented to the annual meeting of the International Studies Association, Atlanta, March 27-31, 1984.

---. "The Coming of World War I: Conflict Reinforcement in the Period 1871-1914." In Theories, Models, and Simulations in International Relations, pp. 419-437. Edited by Michael D. Ward. Boulder, CO: Westview Press, 1985.

Miller, R., and Thorson, Stuart. "Control Concepts in Theories of Governmental Decision-Making: A Structural Approach." In Mathematical Systems in International Relations Research, pp. 60-90. Edited by John V. Gillespie and Dina A. Zinnes. New York: Praeger, 1977.

Miller, Steven E. "Arms Races: The State of the Art." Paper presented to the annual meeting of the American Political Science Association, Chicago, September 3-6, 1987.

Milstein, Jeffrey S. "Soviet and American Influences on the Arab-Israeli Arms Race: A Quantitative Analysis." Peace Science Society (International) Papers. 15 (1970): 6-27.

Milstein, Jeffrey S., and Mitchell, William C. "Computer Simulation of International Processes: The Vietnam War and the Pre-World War I Naval Race." Peace Research Society Papers. 12 (1968): 117-136.

---. "Dynamics of the Vietnam Conflict: A Qualitative Analysis and Predictive Computer Simulation." Peace Research Society Papers. 12 (1968): 163-213.

Mintz, Alex. "A Validation Study of Computer Simulations of the European 'Major Power' System: 1870-1914." Conflict Management and Peace Science. 6:1 (Fall 1981): 19-37.

---. "A Formal Model of Defense Budgeting in Israel." In Theories, Models, and Simulations in International Relations, pp. 287-304. Edited by Michael D. Ward. Boulder, CO: Westview Press, 1985.

---. "The Lavi Decision: A Game Theoretic Analysis." Paper presented to the annual meeting of the International Studies Association, Anaheim, CA, March 25-29, 1986.

Mintz, Alex, and Hicks, A. "Military Keynesianism in the United States, 1949-76: Disaggregating Military Expenditures and Their Determination." American Journal of Sociology. 90 (1984): 411-417.

Moberg, Eric. "Models of International Conflicts and Arms Races." Cooperation and Conflict, Nordic Studies in International Politics. 2 (1966): 80-93.

Moll, Kendall D. The Influence of History Upon Seapower, 1865-1914. Menlo Park, CA: Stanford Research Institute, September 1968.

---. "Politics, Power and Panic: Britain's 1909 Dreadnought 'Gap.'" In The Use of Force, pp. 431-447. Edited by Robert J. Art and Kenneth Waltz. Boston: Little, Brown, 1971.

---. "International Conflict as a Decision System." Journal of Conflict Resolution. 18:4 (December 1974): 555-557.

Moll, Kendall D., and Luebbert, G.M. "Arms Race and Military Expenditure Models." Journal of Conflict Resolution. 24:1 (March 1980): 153-185.

Montrol, E.W., and Badger, W.W. Introduction to Quantitative Aspects of Social Phenomena. New York: Gordon and Breach, 1974.

Morgan, Patrick M. "Arms Races and Strategic Surprise." Paper presented to the annual meeting of the International Studies Association, Los Angeles, March 19-22, 1980.

Morrow, James D. "A Spatial Model of International Conflict." American Political Science Review. 80:4 (December 1986): 1131-1150.

---. "A Limited Information Model of Crisis Bargaining." Paper presented to the annual meeting of the International Studies Association, Washington, DC, April 14-18, 1987.

Moses, Lincoln L. "A Review of Lewis F. Richardson, Arms and Insecurity and Statistics of Deadly Quarrels." Journal of Conflict Resolution. 5:4 (December 1961): 390-394.

Mosley, Hugh G. The Arms Race: Economic and Social Consequences. Lexington, MA: Lexington Books, 1985.

Most, Benjamin A., and Starr, Harvey. "Diffusion, Reinforcement, Geopolitics, and the Spread of War." American Political Science Review. 74:4 (December 1980): 932-946.

Most, Benjamin A.; Schrodt, Philip; Siverson, Randolph M.; and Starr, Harvey. "Border and Alliance Effects in the Diffusion of Major Power Conflict, 1815-1965." Paper presented to the annual meeting of the American Political Science Association, Chicago, September 3-6, 1987.

Moulaert, Frank. "On the Nature and Scope of Complex Conflicts." Paper presented to the annual Peace Science Society (International) Conference, Geneva, September 3-5, 1978.

---. "On the Historical Characterization of Complex Conflicts." Paper presented to the annual Peace Science Society (International) Conference, Berlin, September 2-4, 1979.

---. "On the Nature and Scope of Complex Conflicts." Peace Science Society (International) Papers. 29 (1979-1980): 48-66.

Muncaster, Robert G., and Zinnes, Dina A. "A Model of Inter-nation Hostility Dynamics and War." Conflict Management and Peace Science. 6:2 (Spring 1982-1983): 19-38.

Myrdal, Alva. The Game of Disarmament: How the United States and Russia Run the Arms Race. Manchester, England: Manchester University Press, 1977.

Nacht, Michael L. "The Delicate Balance of Error." Foreign Policy. 19 (Summer 1975): 163-167.

Nash, John F., Jr. "The Bargaining Problem." Econometrica. 18:2 (April 1950): 155-162.

Neimi, G., and Weisberg, Herbert F., eds. Probability Models of Collective Decision Making. Columbus, OH: Charles E. Merrill, 1972.

Newcombe, Alan G., and Andrighetti, Robert. "Nations at Risk: A Prediction of Nations Likely to Be in War in the Years 1974-1978." International Interactions. 3:2 (1977): 135-160.

Newcombe, Alan G., and Wert, James. "The Use of an Inter-nation Tensiometer for the Prediction of War." Peace Science Society (International) Papers. 21 (1973): 73-83.

Nincic, Miroslav. "Fluctuations in Soviet Defense Spending." Journal of Conflict Resolution. 27 (1983): 648-660.

Nincic, Miroslav, and Cusack, Thomas R. "The Political Economy of U.S. Military Spending." Journal of Peace Research. 16:2 (1979): 101-115.

Noel-Baker, Philip. The Arms Race: A Programme for World Disarmament. London: Atlantic Book Publishing Co., 1958.

North, Robert C. "Arms Control and the Dynamics of International Conflict." Stanford Journal of International Studies. 7 (Spring 1972): 96-108.

North, Robert C.; Brody, Richard A.; and Holsti, Ole R. "Some Empirical Data on the Conflict Spiral." Peace Research Society Papers. 1 (1963): 1-14.

Olinick, Michael. "Stable and Unstable Arms Races." In An Introduction to Mathematical Models in the Social and Life Sciences, pp. 20-51. Edited by Michael Olinick. Reading, MA: Addison-Wesley, 1978.

O'Neil, Barry. "The Pattern of Instability among Nations: A Test of Richardson's Theory." General Systems. 15 (1970): 175-181.

Organski, A.F.K., and Kugler, J. The War Ledger. Chicago: University of Chicago Press, 1980.

Organski, A.F.K., and Kugler, J., with Fox, D. "Deterrents, Deterrence and Arms Races." Paper presented to the western meeting of the Peace Science Society (International), Stanford, CA, February 26-27, 1979.

Ostrom, Charles W., Jr. "The Politics of the U.S. Defense Expenditure Policymaking Process: An Empirical Test between an Arms Race Model and an Organizational Politics Model." Ph.D. dissertation, Indiana University, 1975.

---. "Balance of Power and the Maintenance of 'Balance': A Rational Choice Model with Lexical Preferences." In Mathematical Models in International Relations, pp. 318-332. Edited by Dina A. Zinnes and John V. Gillespie. New York: Praeger, 1976.

---. "Evaluating Alternative Foreign Policy Decision-Making Models: An Empirical Test between an Arms Race Model and an Organizational Politics Model." Journal of Conflict Resolution. 21:2 (June 1977): 235-265.

---. "A Reactive Linkage Model of the U.S. Defense Expenditure Policy Making Process." American Political Science Review. 72 (1978): 941-957.

---. Time Series Analysis: Regression Techniques. Beverly Hills, CA: Sage, 1978.

Ostrom, Charles W., Jr., and Job, Brian L. "The President and the Political Use of Force." Paper presented to the annual meeting of the American Political Science Association, Denver, September 2-3, 1982.

Patchen, Martin. "Models of Cooperation and Conflict: A Critical Review." Journal of Conflict Resolution. 14:3 (September 1970): 389-407.

---. "When Do Arms Buildups Lead to Deterrence and When to War?" Paper presented to the World Peace Science Congress, Rotterdam, June 8, 1984.

Pearson, Frederick. "Interaction in an International Political Subsystem: The 'Middle East,' 1963-64." Peace Research Society (International) Papers. 15 (1970): 73-99.

---. "Towards a Regional Model for Foreign Military Intervention and Arms Transfers." Paper presented to the annual meeting of the International Studies Association, Mexico City, April 5-9, 1983.

Phillips, Warren R. "The Dynamics of Behavioral Action and Reaction in International Conflict." Peace Research Society (International) Papers. 17 (1970): 31-46.

---. "Where Have All the Theories Gone?" World Politics. 26:2 (January 1974): 155-188.

---. "Command and Control Tools for Crisis Management." In Mathematical Systems in International Relations Research, pp. 3-26. Edited by John V. Gillespie and Dina A. Zinnes. New York: Praeger, 1977.

Pilisuk, M. "Experimenting with the Arms Race." Journal of Conflict Resolution. 28 (1984): 296-315.

Pitman, G.R., Jr. Arms Races and Stable Deterrence. Los Angeles: UCLA Security Studies Project, 1969.

Polachek, Solomon William. "Dyadic Dispute: An Economic Perspective." Peace Science Society (International) Papers. 28 (1978): 67-80.

Pudaite, Paul R. "The Outbreak of War: A Baysian Inference, Poisson-Markov Process Model." Paper presented to the annual meeting of the International Studies Association, Anaheim, CA, March 25-29, 1986.

Rapoport, Anatol A. "Lewis Fry Richardson's Mathematical Theory of War." Journal of Conflict Resolution. 1:3 (1957): 249-299.

---. Fights, Games and Debates. Ann Arbor: University of Michigan Press, 1961.

---. "The Mathematics of Arms Races." In International Politics and Foreign Policy, pp. 492-503. Edited by James N. Rosenau. New York: Free Press, 1961.

---. "Mathematical Methods in Theories of International Relations: Expectations, Caveats, and Opportunities." In Mathematical Models in International Relations, pp. 10-36. Edited by Dina A. Zinnes and John V. Gillespie. New York: Praeger, 1976.

Rashevsky, N. Mathematical Theory and Human Relations, An Approach to a Mathematical Biology of Social Phenomena. Bloomington, IN: Principa Press, 1947.

---. Mathematical Biology of Social Behavior. Chicago: University of Chicago Press, 1951.

Rasler, Karen. "War, the Great Depression, and American Violence 1890-1970: A Dynamic Analysis." Paper presented to the Conference on Dynamic Models of Conflict, Boulder, CO, October 31 to November 3, 1984.

Rathjens, George R., Jr. "The Dynamics of the Arms Race." Scientific American. 220:4 (April 1969): 15-25.

Rattinger, Hans. "Armaments, Detente, and Bureaucracy: The Case of the Arms Race in Europe." Journal of Conflict Resolution. 19:4 (December 1975): 571-595.

---. "Econometrics and Arms Races: A Critical Review and Some Extensions." European Journal of Political Research. 4 (1976): 421-439.

---. "From War to War to War: Arms Races in the Middle East." International Studies Quarterly. 20 (1976): 501-531.

Reitman, Mark I. "Optimal Level of Nuclear Arms and a Global Strategy of Survival." Paper presented to the annual meeting of the International Studies Association, Anaheim, CA, March 25-29, 1986.

Richardson, Lewis Fry. "Mathematical Psychology of War." N.p.: British Copyright Libraries, 1919.

---. Generalized Foreign Politics. Cambridge: Cambridge University Press, 1939.

---. "The Distribution of Wars in Time." Journal of the Royal Statistical Society. 107:3/4 (1944): 242-250.

---. "Statistics of Deadly Quarrels." In Psychological Factors of Peace and War, pp. 237-255. Edited by T.H. Pear. London: Hutchinson and Co., 1950.

---. "Threats and Security." In Pyschological Factors of Peace and War, pp. 219-235. Edited by T.H. Pear. London: Hutchinson and Co., 1950.

---. "War and Eugenics." Eugenics Review. 42:1 (1950): 25-36.

---. "Could an Arms Race End without Fighting?" Nature. 167 (September 29, 1951): 567.

---. "Is It Possible to Prove Any General Statements About Historical Fact?" British Journal of Sociology. 3 (1952): 77-84.

---. "Mathematics of War and Foreign Policies." In The World of Mathematics, Vol. 2, pp. 1240-1253. Edited by James R. Newman. New York: Simon and Schuster, 1956.

---. Arms and Insecurity: A Mathematical Study of the Causes and Origins of War. Pittsburgh: Boxwood Press, 1960.

---. Statistics of Deadly Quarrels. Chicago: Triangle Press, 1960.

---. "The Problem of Contiguity: An Appendix to Statistics of Deadly Quarrels." General Systems. 6 (1961): 139-187.

Richardson, Stephen A. "Lewis Fry Richardson (1881-1953): A Personal Biography." Journal of Conflict Resolution. 1:3 (1957): 300-307.

Rietvold, Piet. "Analysis of Conflicts and Compromises, Using Qualitative Data." Conflict Management and Peace Science. 7:1 (Fall 1983): 39-63.

Rosenau, James N. "Intellectual Identity and the Study of International Relations, or Coming to Terms with Mathematics as a Tool of Inquiry." In Mathematical Models in International Relations, pp. 3-9. Edited by Dina A. Zinnes and John V. Gillespie. New York: Praeger, 1976.

Ruloff, Dieter. "The Dynamics of Conflict and Cooperation between Nations: A Computer Simulation and Some Results." Journal of Peace Research. 12 (1975): 109-121.

---. Konfliklosung durch Vermittlung. Computersimulation zwischestaatlicher Krisen. Basel: 1975.

---. "Simulation and Gaming: The Analysis of Conflict and Cooperation in the Field of International Relations." In Systems Theory in the Social Sciences. Edited by Bossel et al. Basel: 1975.

---. "Mediation in International Conflicts: A Computer Simulation and Some Results." Peace Science Society (International) Papers. 26 (1976): 22-48.

---. "Cooperation, Integration, and National Characteristics: A Comparative Empirical Analysis Using New Indicators." In Theories, Models, and Simulations in International Relations, pp. 463-482. Edited by Michael D. Ward. Boulder, CO: Westview Press, 1985.

Ruloff, Dieter, and Frei, Daniel. "Forecasting East-West Relations: A Politometric Model." Paper presented to the Conference on Dynamic Models of Conflict, Boulder, CO, October 31 to November 3, 1984.

Rummel, Rudolph J. "Dimensions of Conflict Behavior within and between Nations." General Systems. 8 (1963): 1-50.

Russell, Bertrand. War, the Offspring of Fear. London: Union of Democratic Control, 1914.

Russett, Bruce, ed. What Price Vigilance? The Burdens of National Defense. New Haven: Yale University Press, 1970.

---. "Peace Science and the Continuing Threat of the Bomb." Peace Science Society (International) Papers. 28 (1978): 1-14.

Saaty, Thomas L. "A Model for the Control of Arms." Operations Research. 12:4 (July/August 1964): 586-609.

---. Mathematical Models of Arms Control and Disarmament: Applications of Mathematical Structures in Politics. New York: Wiley and Sons, 1968.

---. "Mathematical Structures Applicable to Multi-party Conflicts." Peace Research Society Papers. 12 (1970): 8-30.

Sampson, Martin W. III. "Policy Zone: Where Policies Work." In Mathematical Models in International Relations, pp. 348-375. Edited by Dina A. Zinnes and John V. Gillespie. New York: Praeger, 1976.

Sandberg, Irwin W. "A Note on Linear Models of Mutual Deterrence." Journal of Peace Science. 2:1 (Spring 1977): 259-260.

---. "Some Qualitative Properties of Nonlinear Richardson-Type Arms Race Models." In Mathematical Systems in International Relations Research, pp. 305-329. Edited by John V. Gillespie and Dina A. Zinnes. New York: Praeger, 1977.

---. "On the Mathematical Theory of Social Processes Characterized by Weak Reciprocity." Journal of Peace Science. 3:1 (Spring 1978): 1-30.

Sanjian, Gregory. "Fuzzy Set Theory and Arms Trade Decision Making." Paper presented to the annual meeting of the International Studies Association, Anaheim, CA, March 25-29, 1986.

Saris, Willem, and Middendorp, Cees. "Arms Races: External Security or Domestic Pressure?" British Journal of Political Science. 10 (1980): 121-128.

Schlenker, B.R., and Bonoma, T.V. "Fun and Games: The Validity of Games for the Study of Conflict." Journal of Conflict Resolution. 22:1 (March 1978): 7-38.

Schrodt, Philip A. "Richardson's N-Nation Model and the Balance of Power." Paper presented to the annual meeting of the International Studies Association, 1975.

---. "The Effects of a Richardson Arms Race on the Distribution of Power in an International System." Ph.D. dissertation, Indiana University, 1976.

---. "Richardson's Model as a Markov Process." In Mathematical Models in International Relations, pp. 156-175. Edited by Dina A. Zinnes and John V. Gillespie. New York: Praeger, 1976.

---. "The Richardson N-Nation Model and the Balance of Power." American Journal of Political Science. 22 (1978): 364-390.

---. "Statistical Problems Associated with the Richardson Arms Race Model." Journal of Peace Research. 3:2 (Fall 1978): 159-172.

---. "Arms and Political Behavior in the Arabian Sea Area: A Statistical Study." Paper presented to the annual meeting of the International Studies Association, Philadelphia, 1981.

---. "Conflict as a Determinant of Territory." Behavioral Science. 26:1 (January 1981): 37-50.

---. "A Mathematical Model of the Persistence of Conflict." International Interactions. 8:4 (1981): 335-348.

---. "Contagion Effects in Arms Transfers: A Statistical Study." Paper presented to the annual meeting of the International Studies Association, Cincinnati, March 25-27, 1982.

---. "Microcomputers in the Study of Politics: Predicting Wars with the Richardson Arms Race Model." Byte. 7:7 (July 1982): 108-134.

---. "Arms Transfers and Supplier/Recepient Influence: A Statistical Study." Paper presented to the annual meeting of the International Studies Association, Mexico City, April 5-9, 1983.

---. "A Model of Sporadic Conflict." In Conflict Processes and the Breakdown of International Systems, pp. 101-118. Edited by Dina A. Zinnes. Denver: Graduate School of International Studies, 1983.

---. "Artificial Intelligence and International Crisis: An Application of Pattern Recognition." Paper presented to the annual meeting of the American Political Science Association, Washington, DC, August 30 to September 2, 1984.

---. "Artificial Intelligence and the State of Mathematical Modeling in International Relations." Paper presented to the Conference on Dynamic Models of Conflict, Boulder, CO, October 31 to November 3, 1984.

---. "Adaptive Precedent-Based Logic and Rational Choice: A Comparison of Two Approaches to the Modeling of International Behavior." In Dynamic Models of International Conflict, pp. 373-400. Edited by Urs Luterbacher and Michael D. Ward. Boulder, CO: Westview Press, 1985.

---. "The Role of Stochastic Models in International Relations Research." In Theories, Models, and Simulations in International Relations, pp. 199-221. Edited by Michael D. Ward. Boulder, CO: Westview Press, 1985.

Schrodt, Philip A.; Gillespie, John V.; and Zinnes, Dina A. "Parameter Estimation by Numerical Minimization Methods." International Interactions. 4:3 (1978): 279-301.

Schrodt, Philip A., and Mintz, Alex. "Patterns of International Interactions in the Middle East." Paper presented to the annual meeting of the International Studies Association, Atlanta, March 27-31, 1984.

Schrodt, Philip A., and Ward, Michael D. "Statistical Inference in Incremental and Difference Equation Formulations." American Journal of Political Science. 25:4 (November 1981): 815-832.

Scott, Andrew M. "CASTELLON: A Computer Simulation of Economic and Political Decision Making." Paper presented to the annual meeting of the International Studies Association, Anaheim, CA, March 25-29, 1986.

Scoville, Herbert, Jr.; Lall, Betty G.; and Hunter, Robert E. The Arms Race: Steps Toward Restraint. New York: Carnegie Endowment for International Peace, 1972.

Seitz, Steven. "Fuzzy Models of Regional Stability and Regional Stabilization." Paper presented to the annual meeting of the International Studies Association, Anaheim, CA, March 25-29, 1986.

---. "Modelling US-USSR Foreign Policy Rivalry." Paper presented to the annual meeting of the International Studies Association, Washington, DC, April 14-18, 1987.

Senghaas, Dieter. Rustung und Militarismus. Frankfurt: Suhrkamp, 1972.

---. "Armament Dynamics and Disarmament." Instant Research on Peace and Violence. 6:1/2 (1976): 3-17.

Shapley, Deborah. "Technology Creep and the Arms Race: ICBM Problem a Sleeper." Science. 201:4361 (September 22, 1978): 1102.

Shisko, Robert. "Defense Budget Interactions Revisited." Rand Paper P-5882. Santa Monica, CA: Rand Corp., 1977.

Siljak, Dragoslav D. "A Competitive Analysis of the Arms Race." Annals of Economic and Social Measurement. 5 (1976): 283-295.

---. "On the Stability of Arms Races." In Mathematical Systems in International Relations Research, pp. 264-296. Edited by John V. Gillespie and Dina A. Zinnes. New York: Praeger, 1977.

Simaan, Marwin, and Cruz, Jose B., Jr. "A Multistage Game Formulation of Arms Race and Control and Its Relationship to Richardson's Model." In Modeling and Simulation, Vol. 4, pp. 149-153. Edited by W.G. Vogt and M.H. Mickle. Pittsburgh: Instrument Society of America, 1973.

---. "Formulation of Richardson's Model of Arms Race from a Differential Game Viewpoint." Review of Economic Studies. 42 (1975): 67-77.

---. "Nash Equilibrium Strategies for the Problem of Armaments Race and Control." Management Science. 22:1 (September 1975): 96-105.

---. "Equilibrium Concepts for Arms Race Problems." In Mathematical Systems in International Relations Research, pp. 342-366. Edited by John V. Gillespie and Dina A. Zinnes. New York: Praeger, 1977.

Simon, Herbert A. "Maintaining the Peace." In Theories, Models, and Simulations in International Relations, pp. 535-547. Edited by Michael D. Ward. Boulder, CO: Westview Press, 1985.

Simowitz, Roslyn L. "Alliances and the Theory of Collective Goods: An Evaluation." In Mathematical Models in International Relations, pp. 333-347. Edited by Dina A. Zinnes and John V. Gillespie. New York: Praeger, 1976.

Singer, J. David. "Threat Perception and the Armament-Tension Dilemma." Journal of Conflict Resolution. 2:1 (1958): 90-105.

---. Quantitative International Politics. New York: Free Press, 1968.

---. "The Outcome of Arms Races: A Policy Problem and a Research Approach." Proceedings of the International Peace Research Association, Third General Conference, 2: 137-146.

---. "Escalation and Control in International Conflict: A Simple Feedback Model." General Systems. 15 (1970): 163-173.

---. "Accounting for International War: The State of the Discipline." Journal of Peace Research. 18:1 (1981): 1-18.

Singer, J. David, and Small, Melvin. The Wages of War 1816-1965: A Statistical Handbook. New York: John Wiley, 1972.

Sisson, Roger L., and Ackoff, Russell L. "Toward a Theory of the Dynamics of Conflict." Peace Research Society Papers. 5 (1966): 183-197.

Siverson, Randolph M., and Duncan, George T. "Stochastic Models of International Alliance Initiation." In Mathematical Models in International Relations, pp. 74-109. Edited by Dina A. Zinnes and John V. Gillespie. New York: Praeger, 1976.

---. "Stochastic Models of the Distribution of Dyadic Warfare in Time." Technical Report No. 120. Pittsburgh: Carnegie-Mellon University, Department of Statistics, 1976.

Smale, S. "The Prisoner's Dilemma and Dynamical Systems Associated to Non-cooperative Games." Economic Journal. 48 (1980): 1617-1634.

Smith, Dale L. "Reagan in Machina: Model-Based Analyses of Recent Changes in American National Security Policy." Paper presented to the annual meeting of the American Political Science Association, Chicago, September 3-6, 1987.

Smith, T.C. "Arms Race Instability and War." Journal of Conflict Resolution. 24:2 (June 1980): 253-284.

Smoker, Paul L. "A Mathematical Study of the Present Arms Race." General Systems. 8 (1963): 51-60.

---. "A Pilot Study of the Present Arms Race." General Systems. 8 (1963): 61-76.

---. "Fear in the Arms Race: A Mathematical Study." Journal of Peace Research. 1 (1964): 55-64.

---. "Sino-Indian Relations: A Study of Trade, Communication, and Defense." Journal of Peace Research. 2 (1964).

---. "The Arms Race: A Wave Model." Peace Research Society Papers. 4 (1965): 151-192.

---. "On Mathematical Models in Arms Races." Journal of Peace Research. 2 (1965): 94-95.

---. "Research Communication on Mathematical Models in Arms Races." Journal of Peace Research. 2:1 (1965): 94-95.

---. "Trade, Defense and the Richardson Theory of Arms Races: A Seven Nation Study." Journal of Peace Research. 4 (1965): 161-176.

---. "A Study of an Arms Race: An Application of the Richardson Model." Lancaster, England: University of Lancaster M.Sc. Thesis, 1966.

---. "The Arms Race as an Open and Closed System." Peace Research Society Papers. 7 (1967): 41-62.

---. "International Processes Simulation." Evanston, IL: Northwestern University Simulated International Processes Project, 1967.

---. "Nation State Escalation and International Integration." Journal of Peace Research. 4 (1967): 61-75.

---. "A Time Series Analysis of Sino-Indian Relations." Journal of Conflict Resolution. 8:2 (June 1969): 172-191.

---. "International Processes Simulation: An Evaluation." Paper prepared for the MSU Events Data Conference, East Lansing, MI, April 15-16, 1970.

---. "International Processes Simulation: A Description." In Experimentation and Simulation in Political Science, pp. 315-365. Edited by J.A. LaPonce and P.L. Smoker. Toronto: Toronto University Press, 1972.

---. "International Relations Simulation." In Simulation in Social and Adminstrative Science, pp. 296-339. Edited by Harold Guetzkow et al. Englewood Cliffs, NJ: Prentice-Hall, 1972.

---. "Towards a Systems Perspective: Directions for the Eighties." In Theories, Models, and Simulations in International Relations, pp. 101-114. Edited by Michael D. Ward. Boulder, CO: Westview Press, 1985.

Snyder, Richard C. "Sciencing as a Policy Science: A Perspective on Guetzkow and Company." In Theories, Models, and Simulations in International Relations, pp. 521-534. Edited by Michael D. Ward. Boulder, CO: Westview Press, 1985.

Sorenson, D.S. "Modeling the Nuclear Arms Race: A Search for Bounded Stability." Journal of Peace Science. 4 (1980): 169-185.

Squires, Michael L. "Three Models of Arms Races." In Mathematical Models in International Relations, pp. 252-273. Edited by Dina A. Zinnes and John V. Gillespie. New York: Praeger, 1976.

---. "Modeling the U.S.-U.S.S.R. Strategic Arms Race." Ph.D. dissertation, Indiana University, 1982.

---. "Modeling the US-USSR Strategic Arms Race." Paper presented to the annual meeting of the International Studies Association, Cincinnati, March 25-27, 1982.

Stahel, Albert A. Die Anwendung der numerischen Mathematik und der Simulationstechnik bei der Darstellung des Ablaufs einer Internationalen Krise. Frankfurt: Huber, 1973.

Starr, Harvey, and Most, Benjamin. "A Return Journey--Richardson, Frontiers, and Wars in the 1946-1965 Era." Journal of Conflict Resolution. 22:3 (September 1978): 441-467.

Steiner, Barry H. Arms Races, Diplomacy, and Recurring Behavior: Lessons from Two Cases. Beverly Hills, CA: Sage, 1973.

Stoll, Richard J. "Two Models of Escalation of Serious Disputes to War, 1816-1976." Paper presented to the annual meeting of the International Studies Association, Los Angeles, March 18-22, 1980.

---. "Let the Researcher Beware: The Use of Richardson Equations to Estimate the Parameters of a Dyadic Arms Acquisition Process." American Journal of Political Science. 26:1 (1982): 77-89.

---. "Nations at the Brink: A Computer Simulation of Government Behavior During Serious Disputes." Simulation and Games. 14:2 (June 1983): 179-200.

---. "Simulating Government Behavior During Disputes." In Theories, Models, and Simulations in International Relations, pp. 501-519. Edited by Michael D. Ward. Boulder, CO: Westview Press, 1985.

---. "System and State in International Politics: A Computer Simulation of Balancing in an Anarchic World." International Studies Quarterly. 31:4 (December 1987): 387-402.

Strauss, Robert P. "An Adaptive Expectations Model of the East-West Arms Race." Peace Research Society Papers. 19 (1972): 29-34.

---. "Interdependent National Budgets: A Model of U.S.-U.S.S.R. Defense Expenditures." In Exploring Competitive Arms Processes, pp. 89-98. Edited by W. Ladd Hollist. New York: Marcel Dekker, 1978.

Stromberg, J.L. The Internal Mechanisms of Defense Budget Processes--Fiscal 1953-1968. Santa Monica, CA: Rand Corp., 1970.

Sylvan, D. "Consequences of Sharp Military Assistance Increases for International Conflict and Cooperation." Journal of Conflict Resolution. 20:4 (December 1976): 609-636.

Taagepera, Rein. "Stockpile-Budget and Ratio Interaction Models for Arms Races." Peace Science Society (International) Papers. 29 (1979-80): 67-78.

Taagepera, Rein; Shiffler, G.; Perkins, T.; and Wagner, D. "Soviet-American and Israeli-Arab Arms Races and the Richardson Model." General Systems. 20 (1975): 151-158.

Taber, G.S. "A 'Learning' Model of War Involvement." Paper presented to the annual meeting of the International Studies Association, Anaheim, CA, March 25-29, 1986.

Tammen, Ronald L. MIRV and the Arms Race: An Interpretation of Defense Strategy. New York: Praeger, 1973.

Tanter, Raymond. "The Policy Relevance of Models in World Politics." Journal of Conflict Resolution. 16 (1972): 555-583.

---. Modeling and Managing International Conflicts. Beverly Hills, CA: Sage, 1974.

Tanter, Raymond, and Potter, William C. "Modelling Alliance Behavior: East-West Conflict Over Berlin." Peace Science Society (International) Papers. 20 (1973): 25-41.

Taylor, J. "Recent Developments in the Lanchester Theory of Combat." In Operations Research '78, pp. 773-806. Edited by K. Haley. Amsterdam: Elsevier North-Holland, 1979.

Thee, Marek. "The Nuclear Arms Race: Trends, Dynamics, and Control." Instant Research on Peace and Violence. 6:1/2 (1976): 18-28.

Thiel, R. Quantitat oder Begriff? der heuristische Gebrauch mathematischer Begriffe in Analyse und Prognose gesellschaftlicher Prozesse. East Berlin: 1967.

Thompson, William R. "Cycles of General, Hegemonic and Global War." Paper presented to the Conference on Dynamic Models of Conflict, Boulder, CO, October 31 to November 3, 1984.

Thompson, William R.; Duval, Robert D.; and Dia, Ahmed. "War, Alliances and Military Expenditures." Journal of Conflict Resolution. 23:4 (December 1979): 629-654.

Thorson, Stuart J.; Anderson, P.A.; Miller, P.M.; Crain, R.C.; and Hainline, M.K. Final Report for the Project for Theoretical Politics. Columbus: Ohio State University, 1977.

Thorson, Stuart J., and Miller, R.A. "Production Systems as Models of Control Structures for Governmental Decision-Making." Paper presented to the National Science Foundation Conference on Control Theory in International Relations Research, Bloomington, IN, April 9-11, 1975.

Valadez, Joseph J. "Exploring International Relations Through Organizational Theories in a Global Systems Simulation: A Training Template." In Theories, Models, and Simulations in International Relations, pp. 249-272. Edited by Michael D. Ward. Boulder, CO: Westview Press, 1985.

Voevodsky, J. "Quantitative Behavior of Warring Nations." Journal of Psychology. 72 (1969): 269-292.

Wagner, D.L.; Perkins, R.T.; and Taagepera, R. "Fitting the Soviet-U.S. Arms Race Data to the Complete Solution to Richardson's Equations." Paper presented to the annual meeting of the Western Regional Peace Science Society, San Diego, February 22, 1974.

---. "Complete Solution to Richardson's Arms Race Equations." Journal of Peace Science. 1:2 (Spring 1975): 159-172.

Wallace, Michael D. "Alliance Polarization, Cross Cutting, and International War, 1815-1964: A Measurement Procedure and Some Preliminary Evidence." Journal of Conflict Resolution. 17:4 (December 1973): 575-604.

---. "Status, Formal Organization, and Arms Levels as Factors Leading to the Onset of War, 1920-1964." In Peace, War, and Numbers. Edited by Bruce M. Russett. Beverly Hills, CA: Sage, 1974.

---. "Arms Races, Domestic Pressures, and Alliances: Some Preliminary Curve Fitting." Paper presented to the world congress of the International Political Science Association, Edinburgh, August 21-25, 1976.

---. "Arms Races and the Balance of Power: A Preliminary Model." Applied Mathematical Modeling. 1:2 (September 1976): 83-92.

---. "Fueling the Arms Race: A Mathematical Model of Nonlinear and Discontinuous Influences on Great Power Arms Expenditures, 1870-1914." In Exploring Competitive Arms Processes, pp. 133-164. Edited by W. Ladd Hollist. New York: Marcel Dekker, 1978.

---. "The Role of Arms Races in the Escalation of Disputes into War: Some New Evidence." Journal of Conflict Resolution. 23:1 (1979): 3-16.

---. "Accounting for Superpower Arms Spending." In Threats, Weapons and Foreign Policy, pp. 259-274. Edited by Pat McGowan and Charles W. Kegley, Jr. Beverly Hills, CA: Sage, 1980.

---. "Some Persisting Findings: A Reply to Professor Weede." Journal of Conflict Resolution. 24:2 (June 1980): 289-292.

---. "Old Nails in New Coffins: The Para Bellum Hypothesis Revisited." Journal of Peace Research. 18:1 (1981): 91-95.

---. "Armaments and Escalation: Two Competing Hypotheses." International Studies Quarterly. 26:1 (1982): 37-56.

---. "Arms Races and Escalation: A Reply to Altfeld." International Studies Quarterly. 27 (1983): 233-235.

Wallace, Michael D., and Wilson, Judy. "Non-linear Arms Race Models." Journal of Peace Research. 15:2 (1978): 175-192.

Ward, Michael D. "Modeling the Dynamics of Cooperation and Conflict in Foreign Policy Behavior: Action, Reaction, and Memory." Evanston, IL: Northwestern University Department of Political Science, 1979.

---. "The Political Economy of Arms Races and International Tension." Conflict Management and Peace Science. 7 (1984): 1-23.

---. "The Differential Paths to Parity: A Study of the Contemporary Arms Race." American Political Science Review. 78:2 (June 1984): 297-317.

---, ed. Theories, Models, and Simulations in International Relations: Essays in Honor of Harold Guetzkow. Boulder, CO: Westview Press, 1985.

---. "Simulation and Stimulation: Four Decades of International Relations Scholarship." In Theories, Models, and Simulations in International Relations, pp. 1-8. Edited by Michael D. Ward. Boulder, CO: Westview Press, 1985.

---. "Dynamic Analysis of Conflict and Cooperation: An Inquiry into the Evolution of International Interactions." In Theories, Models, and Simulations in International Relations, pp. 439-461. Edited by Michael D. Ward. Boulder, CO: Westview Press, 1985.

Ward, Michael D., and House, L. "Probability Models of N-Adic Conflict and Spatial Representation of International Dispute Behavior." Paper presented to the annual meeting of the International Studies Association, Washington, DC, March 5-9, 1985.

Watt, D.W. "The Possibility of a Multilateral Arms Race." International Relations. 2:8 (October 1962).

Weede, Erich. "Arms Races and Escalation: Some Persisting Doubts." Journal of Conflict Resolution. 24:2 (June 1980): 285-287.

---. "Preventing War by Nuclear Deterrence or by Detente." Conflict Management and Peace Science. 6:1 (Fall 1981): 1-18.

---. "Extended Deterrence by Superpower Alliance." Journal of Conflict Resolution. 27 (1983): 231-254.

Weiss, Herbert K. "Lanchester-Type Models of Warfare." In Proceedings of the First International Conference on Operations Research, pp. 82-98. Edited by M. Davies et al. Baltimore: Operations Research Society of America, 1957.

---. "Stochastic Models for the Duration and Magnitude of a 'Deadly Quarrel.'" Operations Research. 11:1 (1963): 101-121.

Westing, Arthur H. "War as a Human Endeavor: High Fatality Wars in the Twentieth Century." Journal of Peace Research. 19:3 (1982): 261-270.

Wilkenfeld, Jonathan. "A Time-Series Perspective on Conflict Behavior in the Middle East." In Sage International Yearbook of Foreign Policy Studies, Vol. 3, pp. 177-212. Edited by Pat McGowan and Charles W. Kegley, Jr. Beverly Hills, CA: Sage, 1975.

Wilkenson, David. Deadly Quarrels: Lewis F. Richardson and the Statistical Study of War. Berkeley: University of California Press, 1980.

Wilson, A. The Bomb and the Computer. New York: Dell, 1968.

Wohlstetter, Albert. "The Delicate Balance of Terror." Foreign Affairs. 37:2 (1959): 211-234.

---. "Is There a Strategic Arms Race?" Foreign Policy. 15 (Summer 1974): 3-20.

---. "Legends of the Strategic Arms Race, Part I: The Driving Engine." Strategic Review. 2 (Fall 1974): 67-92.

---. "Rivals, but No 'Race.'" Foreign Policy. 16 (Fall 1974): 48-81.

---. "Optimal Ways to Confuse Ourselves." Foreign Policy. 20 (Fall 1975): 170-198.

---. "Legends of the Strategic Arms Race, Part II: The Uncontrolled Upward Spiral." Strategic Review. 3 (Winter 1975): 70-86.

---. "Racing Forward, or Ambling Back?" In Defending America, pp. 110-168. Edited by James R. Schlesinger et al. New York: Basic Books, 1977.

Wolfson, Murray. "A Mathematical Model of the Cold War." Peace Research Society Papers. 9 (1968): 107-123.

---. "A Dynamic Model of Present World Conflict." Peace Science Society (International) Papers. 20 (1973): 43-66.

Wright, Quincy. A Study of War. 2 Vols. Chicago: University of Chicago Press, 1942.

---. "The Escalation of International Conflicts." Journal of Conflict Resolution. 9:4 (December 1965): 434-449.

Wright, Steve. "A Multivariate Time Series Analysis of the Northern Irish Conflict, 1969-1976." Peace Science Society (International) Papers. 29 (1979/1980): 28-47.

---. "A Time Series Analysis of the Northern Irish Conflict, 1969-78." Ph.D. thesis, University of Lancaster, 1980.

York, Herbert F. The Race to Oblivion. New York: Simon and Schuster, 1970.

---. "Some Possible Measures for Slowing the Qualitative Arms Race." Proceedings of the Pugwash Conference on Science and World Affairs, pp. 228-235, Oxford, England, September 7-12, 1972.

Zagare, Frank C. "The Pathologies of Unilateral Deterrence." Paper presented to the Conference on Dynamic Models of Conflict, Boulder, CO, October 31 to November 3, 1984.

Zinnes, Dina A. "A Comparison of Hostile Behavior of Decision-Makers in Simulate and Historical Data." World Politics. 18:3 (April 1966): 474-502.

---. Contemporary Research in International Relations. New York: Free Press, 1976.

---. "An Event Model of Conflict Interaction." In Conflict Processes and the Breakdown of International Systems, pp. 119-147. Edited by D.A. Zinnes. Denver: Graduate School of International Studies, 1983.

---. "Modeling International Cooperation." Paper presented to the annual meeting of the International Studies Association, Washington, DC, April 14-18, 1987.

Zinnes, Dina A., and Gillespie, John V. "Analysis of Arms Race Models: USA vs. USSR and NATO vs. WTO." In Modeling and Simulation, Vol. 4, pp. 145-148. Edited by W.G. Vogt and M.H. Mickle. Pittsburgh: Instrument Society of America, 1973.

---. "Introduction to Decision-Making Models." In Mathematical Models in International Relations, pp. 309-317. Edited by Dina A. Zinnes and John V. Gillespie. New York: Praeger, 1976.

---. "Introduction to Probability Models." In Mathematical Models in International Relations, pp. 65-73. Edited by Dina A. Zinnes and John V. Gillespie. New York: Praeger, 1976.

---. "Introduction to Richardson-Type Process Models." In Mathematical Models in International Relations, pp. 179-188. Edited by Dina A. Zinnes and John V. Gillespie. New York: Praeger, 1976.

---, eds. Mathematical Models in International Relations. New York: Praeger, 1976.

Zinnes, Dina A.; Gillespie, John V.; and Rubison, R.M. "A Reinterpretation of the Richardson Arms-Race Model." In Mathematical Models in International Relations, pp. 189-217. Edited by Dina A. Zinnes and John V. Gillespie. New York: Praeger, 1976.

Zinnes, Dina A.; Gillespie, John V.; and Schrodt, Philip A. "An Analysis of a Multistage Game Model of Arms Races." Paper presented to the annual meeting of the American Political Science Association, Chicago, August 28 to September 2, 1974.

---. "The Arab-Israeli Arms Race: An Empirical Examination." Jerusalem Journal of International Relations. 2:1 (Fall 1976): 28-62.

Zinnes, Dina A.; Gillespie, John V.; Schrodt, Philip A.; Tahim, G.S.; and Rubison, R.M. "Arms and Aid: A Differential Game Analysis." In Exploring Competitive Arms Processes, pp. 17-38. Edited by W. Ladd Hollist. New York: Marcel Dekker, 1978.

Zinnes, Dina A.; Gillespie, John V.; Tahim, G.S. "Modeling a Chimera: Balance of Power Revisited." Journal of Peace Science. 3:1 (Spring 1978): 31-44.

---. "The Richardson Arms Race Model." UMAP Unit 308. Newton, MA: EDC/UMAP, 1978.

---. "The Richardson Arms Race Model." Bloomington: Indiana University Center for International Policy Studies, 1978.

---. "Transforming an International System." Journal of Conflict Resolution. 22 (1978): 547-564.

Zinnes, Dina A.; Hill, Barbara J.; Jones, David L.; and Majeski, Stephen J. "Modeling Precrisis Interactions." In Missing Elements in Political Inquiry, pp. 117-140. Edited by Judith A. Gillespie and Dina A. Zinnes. Beverly Hills, CA: Sage, 1982.

Zinnes, Dina A., and Muncaster, Robert. "Systems Analysis and the Potential for War." Paper presented to the annual meeting of the International Studies Association, Atlanta, March 27-31, 1984.

Other Sources Consulted

Abelson, R. "The Structure of Belief Systems." In Computer Models of Thought and Language. Edited by Roger Shank and Kenneth Colby. San Francisco: Freeman, 1973.

Agnew, H.M. "Technological Innovation: A Necessary Deterrent or Provocation?" Air Force and Space Digest. 50 (May 1967): 66-70.

Aleksandrov, A.D.; Kolmogorov, A.N.; and Lavrentev, M.A., eds. Mathematics: Its Content, Methods, and Meaning, 2d ed. 3 volumes. Cambridge, MA: MIT Press, 1984.

Alexander, Christopher. Notes on the Synthesis of Form. Cambridge, MA: Harvard University Press, 1964.

Aliano, Richard. American Defense Policy from Eisenhower to Kennedy: The Politics of Changing Military Requirements 1957-1961. Athens: Ohio University Press, 1975.

Alker, Hayward R., Jr. "Can the End of 'Power Politics' Be Part of the Concepts with Which Its Story Is Told?" Paper presented at the annual meeting of the American Political Science Association, Washington, DC, September 2, 1977.

---. "From Political Cybernetics to Global Modeling." Paper presented to the annual meeting of the American Political Science Association, New York, August 31 to September 3, 1978.

Alker, Hayward R., Jr., and Christensen, Cheryl. "From Causal Modeling to Artificial Intelligence: The Evolution of a UN Peace-Making Simulation." In Experimentation and Simulation in Political Science, pp. 177-224. Edited by J.A. LaPonce and Paul Smoker. Toronto: Toronto University Press, 1972.

Allen, Thomas B. War Games: The Secret World of the Creators, Players and Policy Makers Rehearsing World War III Today. New York: McGraw-Hill, 1987.

Allison, Graham T. The Essence of Decision: Explaining the Cuban Missile Crisis. Boston: Little, Brown and Co., 1971.

Anderson, Paul A., and Thorson, Stuart J. "Artificial Intelligence Based Simulation of Foreign Policy Decision-Making." Behavioral Science. 27 (April 1982): 176-193.

Anderson, T.W. The Statistical Analysis of Time Series. New York: Wiley, 1971.

Ando, Albert; Fisher, Franklin M.; and Simon, Herbert A. Essays on the Structure of Social Science Models. Cambridge, MA: MIT Press, 1963.

Armacost, Michael H. The Politics of Weapons Innovation: The Thor-Jupiter Controversy. New York: Columbia University Press, 1969.

Art, R.J. The TFX Decision: McNamara and the Military. Boston: Little, Brown and Co., 1968.

---. "Bureaucratic Politics and American Foreign Policy: A Critique." Policy Sciences. 4 (December 1973): 467-490.

Art, R.J., and Waltz, Kenneth W., eds. The Uses of Force. Boston: Little, Brown and Co., 1971.

Ashby, W.R. An Introduction to Cybernetics. New York: John Wiley and Sons, 1958.

---. Design for a Brain: The Origin of Adaptive Behavior. New York: John Wiley and Sons, 1960.

Ashley, Richard K. "Noticing Pre-paradigmatic Progress." In In Search of Global Patterns, pp. 150-158. Edited by James N. Rosenau. New York: Free Press, 1976.

Banerjee, Sanjoy. "Technical Knowledge in a Political Economy Context: An Artificial Intelligence Based Model." Paper presented to the annual meeting of the International Studies Association, Anaheim, CA, March 25-29, 1986.

Barr, Avron. "Meta-Knowledge and Cognition." Proceedings of the sixth International Joint Conference on Artificial Intelligence, Tokyo, August 20-23, 1979, Vol. 1, pp. 31-33.

Barr, Avron, and Feigenbaum, Edward A. The Handbook of Artificial Intelligence. 3 vols. Stanford, CA: HeurisTech Press, 1982.

Bateson, Gregory. "Culture Contact and Schismogenisis." Man. 35 (1935): 173-183.

---. Steps to an Ecology of Mind. New York: Ballantine Books, 1972.

Beal, Richard S. "Systems-Cybernetic Analysis and the Study of International Politics." Paper presented to the annual meeting of the American Political Science Association, New York, August 31 to September 3, 1978.

Beard, Edward. Developing the ICBM: A Study in Bureaucratic Politics. New York: Columbia University Press, 1976.

Bennett, James P. "Data Stories: Learning about Learning from the U.S. Experience in Vietnam." In Foreign Policy Decision Making: Perception, Cognition, and Artificial Intelligence, pp. 227-279. Edited by D.A. Sylvan and S. Chan. New York: Praeger, 1984.

---. "On the Felicity of Assured Destruction: Building Expert Systems in Areas Lacking Knowledge." Paper presented to the annual meeting of the International Studies Association, Washington, DC, March 5-9, 1985.

Bennett, Scott, and Schrodt, Philip A. "Linear Event Patterns in WEIS Data." Paper presented to the annual meeting of the American Political Science Association, Chicago, September 3-6, 1987.

Berkeley, E.C., and Bobrow, Daniel G. The Programming Language LISP. Cambridge, MA: Information International, 1964.

---. The Programming Language LISP: Its Operation and Applications. Cambridge, MA: MIT Press, 1966.

Blakley, Daniel L.; Cohen, Kalman J.; Lewin, Arie Y.; and Morey, Richard C. "Assessing Defense Procurement Policies." Simulation. 38:3 (March 1982): 75-83.

Blalock, H.M., Jr. Conceptualization and Measurement in the Social Sciences. Beverly Hills, CA: Sage, 1982.

Bluhm, W.T. "Hybris and Aggression: A Critique of Lebow's 'Paranoia of the Powerful' and an Alternative Theory." PS. 17 (1984): 586-591.

Bobrow, Daniel G., and Collins, A., eds. Representation and Understanding. New York: Academic Press, 1975.

Bobrow, Daniel G., and Winograd, Terry. "An Overview of KRL, a Knowledge Representation Language." Cognitive Science. 1:1 (1977): 3-46.

Bobrow, Davis B., ed. Weapons Systems Decisions: Perspectives on Continental Defense. New York: Praeger, 1969.

Bobrow, Davis B.; Sylvan, Donald A.; and Ripley, Brian. "Japanese Supply Security: A Computational Model." Paper presented to the annual meeting of the International Studies Association, Anaheim, CA, March 25-29, 1986.

Boden, Margaret. Artificial Intelligence and Natural Man. New York: Basic Books, 1977.

Bonham, G.M. "Simulating International Disarmament Negotiations." Journal of Conflict Resolution. 15:9 (1971): 299-316.

Box, G.E.P., and Jenkins, G.M. Time Series Analysis: Forecasting and Control. San Francisco: Holden-Day, 1970.

Boydston, Robert. "Research on Non-violent Strategies: Artificial Intelligence as a Methodological Tool." Paper presented to the annual meeting of the International Studies Association, Washington, DC, March 5-9, 1985.

Braybrooke, David, and Lindbloom, Charles E. A Strategy of Decision. New York: Free Press, 1963.

Brewer, Garry D., and Shubik, Martin. The War Game: A Critique of Military Problem Solving. Cambridge, MA: Harvard University Press, 1979.

Brodie, Bernard, ed. The Absolute Weapon: Atomic Power and World Order. New York: Harcourt, Brace, 1946.

---. Strategy in the Missile Age. Princeton, NJ: Princeton University Press, 1959.

Brodie, Bernard, and Brodie, Fawn M. From Crossbows to H-Bomb: The Evolution of the Weapons and Tactics of Warfare. Bloomington: Indiana University Press, 1973.

Brody, Richard A. "Varieties of Simulations in International Relations Research." In Simulations in International Relations, pp. 190-223. Edited by Harold Guetzkow et al. Englewood Cliffs, NJ: Prentice-Hall, 1963.

Brown, Michael E. "The Organizational Sources of Weapon Acquisition Policy." Paper presented to the annual meeting of the American Political Science Association, Chicago, September 3-6, 1987.

Buchanan, Bruce G. "Research on Expert Systems." Report Number STAN-CS-81-837. Palo Alto, CA: Stanford University, Department of Computer Science, 1981.

Buckley, W., ed. Modern Systems Theory for the Behavioral Sciences. Chicago: Aldine, 1969.

Bush, Vannevar. Modern Arms and Free Men: A Discussion of the Role of Science in Preserving Democracy. New York: Simon and Schuster, 1949.

Butterfield, Herbert. The Origins of Modern Science, 1300-1800. New York: Free Press, 1957.

Carbonell, Jamie G. "POLITICS: Automated Ideological Reasoning." Cognitive Science. 2 (1978): 27-51.

---. "The Counterplanning Process: A Model of Decision-Making in Adverse Situations." Pittsburgh: Carnegie-Mellon University, Department of Computer Science, February 1979.

---. "POLITICS." In Inside Computer Understanding, pp. 259-307. Edited by R. Schank and C. Riesbeck. Hillsdale, NJ: Lawrence Erlbaum, 1981.

Charniak, Eugene; Riesbeck, Christopher K.; and McDermott, Drew V. Artificial Intelligence Programming. Hillsdale, NJ: Lawrence Erlbaum, 1980.

Charniak, Eugene, and Wilks, Y. Computational Semantics: An Introduction to Artificial Intelligence and Natural Language Comprehension. Amsterdam: North-Holland Publishers, 1976.

Chomsky, Noam. "Deep Structure, Surface Structure, and Semantic Interpretation." In Semantics: An Interdisciplinary Reader, pp. 183-216. Edited by Danny D. Steinberg and Leon A. Jakobvits. Cambridge, England: Cambridge University Press, 1972.

Cimbala, Stephen J. Artificial Intelligence and National Security. Lexington, MA: Lexington Books, 1986.

Clark, Thomas D., Jr.; McCollough, Brian J.; and Templin, Ralph J. "A Conceptual Model of the Effects of Department of Defense Realignments." Behavioral Science. 25 (1980): 149-160.

Clemens, Walter C., Jr. The Arms Race and Sino-Soviet Relations. Stanford, CA: Stanford University Press, 1968.

---. "The Ecology of Weaponry." The Bulletin of the Atomic Scientists. 26 (September 1970): 27-31.

Cockburn, Andrew. The Threat: Inside the Soviet Military Machine. New York: Vintage, 1984.

Colby, Kennth M. Artificial Paranoia. New York: Pergamon Press, 1975.

Cooling, B. Franklin III. "The Formative Years of the Naval Industrial Complex: Their Meaning for Studies of Institutions Today." Naval War College Review. 27 (March/April 1975): 53-62.

Coplin, W.D., ed. Simulation Models of the Decision-Maker's Environment. Chicago: Markham Publishing Co., 1968.

Coplin, W.D.; O'Leary, M.K.; and Mills, S.L. "An Analytical Description of PRINCE." Syracuse, NY: Syracuse University (mimeo), 1971.

Corkill, Daniel D. "Hierarchical Planning in a Distributed Environment." Proceedings of the sixth International Joint Conference on Artificial Intelligence, Tokyo, August 20-23, 1979, Vol. 1, pp. 168-175.

Cortes, F.; Prezeworski, A.; and Sprague, J. Systems Analysis for Social Scientists. New York: John Wiley and Sons, 1974.

Coulam, R.F. Illusions of Choice: The F-111 and the Problems of Weapons Acquisition Reform. Princeton, NJ: Princeton University Press, 1977.

Crecine, J.P. Governmental Problem Solving: A Computer Simulation of Municipal Budgeting. Chicago: Rand McNally, 1969.

Cyert, R., and March, James G. A Behavioral Theory of the Firm. Englewood Cliffs, NJ: Prentice-Hall, 1963.

Davis, D.A.; Dempster, M.A.H.; and Wildavsky, A. "A Theory of the Budgetary Process." American Political Science Review. 60 (1966): 529-547.

Davis, Randall. "Applications of Meta-Level Knowledge to the Construction, Maintenance and Use of Large Knowledge Bases." Ph.D. dissertation, Stanford University, Department of Computer Science, 1976.

Davis, Randall, and Buchanan, Bruce. "Meta-level Knowledge: Overview and Applications." Proceedings of the International Joint Conference on Artificial Intelligence, Cambridge, MA, August 22-25, 1977.

Davis, Randall; Buchanan, Bruce; and Shortliffe, E. "Production Rules as a Representation for a Knowledge-Based Consultation System." Artificial Intelligence. 8 (1977): 15-45.

Davis, Robert H. "Arms Control Simulation: The Search for an Acceptable Method." Journal of Conflict Resolution. 7:3 (1963): 590-602.

Destler, I.M. Presidents, Bureaucracies and Foreign Policy: The Politics of Organizational Reform. Princeton, NJ: Princeton University Press, 1971.

Deutsch, Karl. The Nerves of Government. New York: Free Press, 1963.

Deutsch, Karl W.; Fritsch, Bruno; Juguaribe, Helio; and Markovits, A.S., eds. Problems of World Modeling. Cambridge, MA: Ballinger Publishing Co., 1977.

Dingman, Roger. "Strategic Planning and the Policy Process: American Plans for War in East Asia, 1945-1950." Naval War College Review. 32:6 (November/December 1979): 4-21.

Dreyfus, Hubert L. "From Micro-Worlds to Knowledge Representation: AI at an Impasse." In Mind Design: Philosophy, Psychology, Artificial Intelligence, pp. 161-204. Edited by John Haugeland. Cambridge, MA: MIT Press, 1981.

Duda, R.; Gaschnig, J.; and Hart, P. "Model Design in the PROSPECTOR Consultant System for Mineral Exploration." In Expert Systems in the Micro-Electronic Age, pp. 153-167. Edited by Donald Michie. Edinburgh: Edinburgh University Press, 1979.

Duffy, Gavan, and Mallery, John. "RELATUS: An Artificial Intelligence Tool for Natural Language Modeling." Paper presented to the annual meeting of the International Studies Association, Anaheim, CA, March 25-29, 1986.

Earle, E.M., ed. The Makers of Modern Strategy: Military Thought from Machiavelli to Hitler. New York: Atheneum, 1971.

Easton, David. The Political System: An Inquiry into the State of Political Science, 2d ed. New York: Alfred A. Knopf, 1971.

Eban, Abba. The New Diplomacy: International Affairs in the Modern Age. New York: Random House, 1983.

Edelson, E. "Expert Systems--Computers That Think Like People." Popular Science. 221 (September 1982): 48-52.

Engleman, C.; Berg, Charles H.; and Bischoff, Miriam. "KNOBS: An Experimental Knowledge Based Tactical Air Mission Planning System." Proceedings of the sixth International Joint Conference on Artificial Intelligence, Tokyo, August 20-23, 1979, Vol. 1, pp. 247-249.

Ensign, Margee M. "Artificial Intelligence-Based Simulations in International Relations." Paper presented to the annual meeting of the International Studies Association, Anaheim, CA, March 26-29, 1986.

Etcheson, Craig. "Expert Systems for Country Risk Analysis." Paper presented to the annual meeting of the International Studies Association, Atlanta, March 27 to April 1, 1984.

Etzioni, Amitai. Modern Organizations. Englewood Cliffs, NJ: Prentice-Hall, 1964.

Feigenbaum, Edward A. "Themes and Case Studies of Knowledge Engineering." In Expert Systems in the Micro-Electronic Age, pp. 3-25. Edited by Donald Michie. Edinburgh: Edinburgh University Press, 1979.

Findler, N.V., ed. Associative Networks: Representation and Use of Knowledge by Computers. New York: Academic Press, 1979.

Finkelstein, Z., and Smith, N. "False Dichotomies in the Defense Debate." Parameters. 17:1 (1987): 89-93.

Finley, M.I. Economy and Society in Ancient Greece. New York: Viking Press, 1953.

Fox, Mark S. "On Inheritance in Knowledge Representation." Proceedings of the sixth International Joint Conference on Artificial Intelligence, Tokyo, August 20-23, 1979, Vol. 1, pp. 282-284.

Gilbert, Felix. The End of the European Era, 1890 to the Present. New York: Norton, 1970.

Gilpin, Robert. American Scientists and Nuclear Weapons Policy. Princeton, NJ: Princeton University Press, 1962.

---. War and Change in World Politics. New York: Cambridge University Press, 1981.

Goldberg, A., and Kay, A. SMALLTALK-72 Instruction Manual. SSL 76-6. Palo Alto, CA: Xerox Palo Alto Research Center, 1976.

Gorden, M. "Burdens for the Designer of a Computer Simulation of International Relations: The Case of TEMPER." In Computers and the Policy-Making Community, pp. 222-245. Edited by Daniel G. Bobrow and J.V. Schwartz. Englewood Cliffs, NJ: Prentice-Hall, 1968.

Gorlin, D.; Hayes-Roth, Frederick; Rosenschein, S.; and Sowizral, Henry. Programming in ROSIE: An Introduction by Means of Examples. N-1646-ARPA. Santa Monica, CA: Rand Corp., 1982.

---. Rationale and Motivation for ROSIE. N-1648-ARPA. Santa Monica, CA: Rand Corp., 1982.

---. The ROSIE Language Reference Manual. N-1657-APRA. Santa Monica, CA: Rand Corp., 1982.

Granger, C.W.J., and Newbold, P. Forecasting Economic Time Series. New York: Academic Press, 1977.

Gray, Robert C. "Learning from History: Case Studies in the Weapons Acquisition Process." World Politics. 31:3 (April 1979): 457-470.

Greenwood, T. Making the MIRV: A Study in Defense Decision-Making. Cambridge, MA: Ballinger, 1975.

Guetzkow, Harold, ed. Simulation in the Social Sciences: Readings. Englewood Cliffs, NJ: Prentice-Hall, 1962.

Guetzkow, Harold; Alger, Chadwick F.; Brodie, Richard A.; Noel, R.C.; and Snyder, R.C. Simulation in International Relations: Developments for Research and Teaching. Englewood Cliffs, NJ: Prentice-Hall, 1963.

Guetzkow, Harold; Kotler, Philip; and Schultz, Randall L., eds. Simulation in Social and Administrative Science: Overviews and Case-Examples. Englewood Cliffs, NJ: Prentice-Hall, 1972.

Halperin, Morton H. "The Decision to Deploy the ABM: Bureaucratic and Domestic Politics in the Johnson Administration." World Politics. 25 (October 1972): 62-95.

---. Bureaucratic Politics and Foreign Policy. Washington, DC: Brookings Institution, 1974.

Harold, Robert C., and Mahoney, Shane E. "Military Hardward Procurement: Some Comparative Observations on Soviet and American Policy Processes." Comparative Politics. 6 (July 1974): 571-599.

Haugeland, J., ed. Mind Design: Philosophy, Psychology, and Artificial Intelligence. Cambridge, MA: MIT Press, 1981.

Hayes-Roth, Barbara; Hayes-Roth, Frederick; Rosenschein, Stan; and Cammarata, Stephanie. "Modeling Planning as an Incremental, Opportunisitic Process." Proceedings of the sixth International Joint Conference on Artificial Intelligence, Tokyo, August 20-23, 1979, Vol. 1, pp. 375-383.

Hayes-Roth, Frederick. "A Tutorial on Expert Systems: Putting Knowledge to Work." Proceedings of the seventh International Joint Conference on Artificial Intelligence, University of British Columbia, Vancouver, August 24-28, 1981.

Hayes-Roth, Frederick; Waterman, Donald A.; and Lenat, Douglas, eds. Building Expert Systems. Reading, MA: Addison-Wesley, 1983.

Hegel, Georg W.F. The Philosophy of Right. Translated by T.M. Knox. Chicago: Encyclopaedia Britannica, 1952.

Heims, Steve J. John von Neumann and Norbert Weiner: From Mathematics to the Technologies of Life and Death. Cambridge, MA: MIT Press, 1980.

Heroditus. The Histories of Heroditus. Translated by George Rawlinson. Chicago: Encyclopaedia Britannica, 1952.

Herrmann, Richard. "The Empirical Challenge of the Cognitive Revolution: A Strategy for Drawing Inferences About Perceptions." International Studies Quarterly. 32:2 (June 1988): 175-203.

Hewitt, C. "Viewing Control Structures as Patterns of Passing Messages." Artificial Intelligence. 8:3 (1977): 323-364.

Hilton, Gorden. "An Appraisal of the Philosophy of Science and Research Design Involved in the Dimensionality of Nations Project." In Quantitative International Politics: An Appraisal, pp. 155-175. Edited by Francis W. Hoole and Dina A. Zinnes. New York: Praeger, 1976.

Hitch, C.J., and McKean, R.N. The Economics of Defense in the Nuclear Age. Cambridge, MA: Harvard University Press, 1963.

Hobbes, Thomas. Leviathan. Edited by Nelle Fuller. Chicago: Encyclopaedia Britannica, 1952.

Holland, Lauren. "Explaining Weapons Procurement: Weaving Old Conceptual Threads into New Theory." Paper presented to the annual meeting of the American Political Science Association, Chicago, September 3-6, 1987.

Holley, I.B., Jr. Ideas and Weapons: Exploitation of the Aerial Weapon by the U.S. During World War I--A Study in the Relationship of Technological Advance, Military Doctrine and the Development of Weapons. New Haven: Yale University Press, 1953.

Hoole, Francis W., and Zinnes, Dina A., eds. Quantitative International Politics: An Appraisal. New York: Praeger, 1976.

Hudson, Valerie M. "Some Issues in the Application of Artificial Intelligence Methodology to the Study of International Relations." Paper presented to the annual meeting of the International Studies Association, Washington, DC, March 5-9, 1985.

Huntington, Samuel P. The Common Defense: Strategic Programs in National Politics. New York: Columbia University Press, 1961.

---. "Interservice Competition and the Political Roles of the Armed Services." In Total War and Cold War: Problems in Civilian Control of the Military, pp. 178-210. Edited by Harry L. Coles. Columbus: Ohio State University Press, 1962.

Hurwitz, Roger; Mallery, John; Alker, Hayward R., Jr.; and Duffy, Gavan. "Anarchy or Community? A Comparative Study of Developmental Patterns in SPD Games." Paper presented to the annual meeting of the International Studies Association, Anaheim, CA, March 25-29, 1986.

IQLISP Reference Manual. Seattle: Integral Quality, 1983.

Jackson, P.C. An Introduction to Artificial Intelligence. New York: Mason Lipscomb, 1974.

James, F. "Function Minimization." Proceedings of the CERN Computing and Data Processing School, Pertisau, Austria, September 10-24, 1972.

---. "Interpetation of the Errors on Parameters as Given by MINUIT." Geneva: CERN Computer Center Program Library, January 1978.

James, F., and Roos, M. MINUIT. Geneva: CERN Computer Center Program Library, September 1977.

Job, Brian, and Johnson, Douglas. "A Formal Decision-Making Model of U.S. Intervention: The U.S. and the Dominican Republic." Paper presented to the annual meeting of the International Studies Association, Anaheim, CA, March 25-29, 1986.

Job, Brian; Johnson, Douglas; and Selbin, Eric. "United States Intervention in Central America: A Model and Application to the Dominican Republic, 1959-1965." Paper presented to the annual meeting of the American Political Science Association, Chicago, September 3-6, 1987.

Johnston, J. Econometric Methods, 2d ed. New York: McGraw-Hill, 1972.

Kahn, K.M. DIRECTOR Guide. AI Memo 482-B. Cambridge, MA: MIT Artificial Intelligence Lab, 1979.

Kantor, Arnold, and Thorson, Stuart J. "The Weapons Procurement Process: Choosing among Competing Theories." Public Policy. 20 (Fall 1972): 479-524.

Kaplan, Abraham. The Conduct of Inquiry: Methodology for Behavioral Science. New York: Thomas Y. Crowell, 1964.

Kaplowitz, Noel. "Psychopolitical Dimensions of International Relations: The Reciprocal Effects of Conflict Strategies." International Studies Quarterly. 28:4 (December 1984): 373-406.

Kennedy, Paul. The Rise of Anglo-German Antagonism 1860-1914. London: George Allen and Unwin, 1980.

King, M., and Hayes, P. "An Introduction to Programming in LISP." In Computational Semantics, pp. 235-282. Edited by E. Charniak and Y. Wilks. Amsterdam: North-Holland Publishing, 1976.

Klahr, Philip, and Faught, William S. "Knowledge-Based Simulation." Proceedings of the first annual meeting of the American Association for Artificial Intelligence, Stanford, CA, August 18-21, 1980, pp. 181-183.

Klahr, Philip; McArthur, David; Narain, Sanjai; and Best, Eric. "SWIRL: Simulating Warfare in the ROSS Language." Report N-1885-AF. Santa Monica, CA: Rand Corp., 1982.

Kline, R.R. "The Beginnings of Modern Management: Alexander and Darius." Journal of Systems Management. 39 (1985): 22-26.

Klockner, John D., and Bennett, W. Lance. "Inferring Belief System Structure and Operation from Resource Planning Requirements." Paper presented to the annual meeting of the International Studies Association, Washington, DC, April 14-18, 1987.

Kmenta, J. Elements of Econometrics. New York: Macmillan, 1971.

Kreutzer, Wolfgang. "Patterns of Modeling: Toward a Conceptual Basis for Discrete Event Simulation." SIMULETTER (SIGM-ACM). 11:3-4 (Spring/Summer 1980): 7-23.

Kuhn, Thomas S. The Structure of Scientific Revolutions, 2d ed. Chicago: University of Chicago Press, 1970.

Lakatos, Irme. "Falsification and the Methodology of Scientific Research Programmes." In Criticism and the Growth of Knowledge, pp. 91-196. Edited by Irme Lakatos and Alan Musgrave. London: Cambridge University Press, 1970.

Lane, Ruth. "Artificial Intelligence and the Political Construction of Reality: The Case of James E. Carter." Paper presented to the annual meeting of the American Political Science Association, September 5-9, 1986.

Lasswell, Harold D. Politics: Who Gets What, When, How. New York: Peter Smith, 1950.

Lebow, R.N. "The Paranoia of the Powerful: Thucydides and World War III." PS. 17 (1984): 10-17.

---. "Thucydides and Aggression: A Reply to Professor Bluhm." PS. 17 (1984): 591-594.

Lenat, Douglas B.; Hayes-Roth, Frederick; and Klahr, Philip. "Cognitive Economy in Artificial Intelligence Systems." Proceedings of the sixth International Joint Conference on Artificial Intelligence, Tokyo, August 20-23, 1979, Vol. 1, pp. 531-536.

Levitt, T.S.; Kirby, R.L.; and Muller, H. "Model-Based System for Force Structure Analysis." Paper presented at the ninth annual Technical Symposium on Optics and Electro-Optics, Arlington, VA, April 8-12, 1985.

Liddell-Hart, B.H. Strategy. New York: Signet, 1967.

Lincoln, James B. "Weapons Systems Acquisition Process." Military Review. 51 (August 1971): 40-52.

Lindbloom, Charles E. "The Science of Muddling Through." Public Administration Review. 19 (Spring 1959): 79-88.

Long, Franklin A., and Reppy, Judith, eds. The Genesis of New Weapons: Decision-Making for Military R&D. New York: Pergamon Press, 1980.

Lucas, John J., and Wait, John V. DARE-P User's Manual. N.p.: (mimeo), 1974.

Lucas, William A., and Dawson, Raymond H. The Organizational Politics of Defense. Pittsburgh: Center for International Studies, 1974.

Luce, R. Duncan, and Raiffa, Howard. Games and Decisions: Introduction and Critical Survey. New York: John Wiley and Sons, 1957.

Luttwak, Edward. Strategy and Politics. New Brunswick, NJ: Transaction, 1980.

McArthur, David, and Klahr, Philip. "The ROSS Language Manual." Report N-1854-AF. Santa Monica, CA: Rand Corp., 1982.

McCarthy, J.; Abrahams, P.W.; Edwards, D.J.; Hart, T.P.; and Levin, M.I. LISP 1.5 Programmer's Manual. Cambridge, MA: MIT Press, 1962.

McClelland, P.D. Causal Explanation: Model-Building in History, Economics and the New Economic History. Ithaca, NY: Cornell University Press, 1975.

McClendon, R. Earl. The Question of Autonomy for the United States Air Arm, 1907-1945. Montgomery, AL: Maxwell Air Force Base, Air University Library, 1950.

McDermott, Drew. "Artificial Intelligence Meets Natural Stupidity." In Mind Design: Philosophy, Psychology, Artificial Intelligence, pp. 143-160. Edited by John Haugeland. Cambridge, MA: MIT Press, 1981.

McDermott, John. "Learning to Use Analogies." Proceedings of the sixth International Joint Conference on Artificial Intelligence, Tokyo, August 20-23, 1979, Vol. 1, pp. 568-576.

Maines, David R. "Social Organization and Social Structure in Symbolic Interactionist Thought." Annual Review of Sociology. 3 (1977): 235-239.

Maisel, Herbert, and Gnuglino, Giuliano. Simulation of Discrete Stochastic Systems. Chicago: Science Research Associates, 1972.

Majeski, Stephen, and Sylvan, David. "Decision Rules for War Involvement: Vietnam Revisited." Paper presented to the annual meeting of the International Studies Association, Washington, DC, March 5-9, 1985.

---. "A Computational Model of the U.S. Foreign Policy Recommendation Process: Vietnam and Laos in 1961." Paper presented to the annual meeting of the American Political Science Association, Chicago, September 3-6, 1987.

Mallery, John C.; Duffy, Gavin; and Hurwitz, Roger. "Computing Strategic Language: Natural Language Models of Belief and Intention." Paper presented to the annual meeting of the International Studies Association, Washington, DC, April 14-18, 1987.

Mallery, John C., and Hurwitz, Roger. "Analogy and Precedent in Strategic Decision-Making: A Computational Approach." Paper presented to the annual meeting of the American Political Science Association, Chicago, September 3-6, 1987.

March, James G. Handbook of Organizations. Chicago: Rand McNally, 1965.

March, James G., and Simon, Herbert A. Organizations. New York: John Wiley and Sons, 1958.

Marcuse, Herbert. Negations. Boston: Beacon Press, 1968.

Martins, Gary. "Better Simulation Models for Decision Support." Data Base. 12:1-2 (Fall 1980): 47-54.

Mayo, R. Britton. Corporate Planning and Modeling with SIMPLAN. Reading, MA: Addison-Wesley, 1979.

Medalian, Jonathan E. "Congress and the Political Guidance of Weapons Procurement." Naval War College Review. 27 (Fall 1975): 12-31.

Mefford, Dwain. "Case Studies and Case Study Systems." Paper presented to the annual meeting of the International Studies Association, Los Angeles, March 19-22, 1980.

---. "Formulating Foreign Policy on the Basis of Historical Analogies: An Application of Developments in Artificial Intelligence." Paper presented to the annual meeting of the International Studies Association, Atlanta, March 27-31, 1984.

---. "Using the Past to Create the Future: The Role of Cognition and Memory in Crisis and War." Paper presented to the annual meeting of the International Studies Association, Atlanta, March 27-31, 1984.

---. "Behind Every Solution Is a Story: Story Grammars and Extended Game Theory." Paper presented to the annual meeting of the International Studies Association, Washington, DC, March 5-9, 1985.

---. "Changes in Foreign Policy Across Time: The Logical Analysis of a Succession of Decision Problems Using Logic Programming." In Dynamic Models of International Conflict, pp. 401-423. Edited by Urs Luterbacher and Michael D. Ward. Boulder, CO: Lynne Reinner Publishers, 1985.

---. "The Postulate of Rationality: A Conceptual Foundation for Artificial Intelligence and the Human Sciences." Paper presented to the annual meeting of the International Studies Association, Washington, DC, April 14-18, 1987.

---. "Case Based Reasoning: Artificial Intelligence Applied to American Foreign Policy." Paper presented to the annual meeting of the American Political Science Association, Chicago, September 3-6, 1987.

Mesarovic, M.D.; Macko, D.; and Takahara, Y. Theory of Hierarchical, Multilevel Systems. New York: Academic Press, 1971.

Michie, Donald. "Expert Systems." The Computer Journal. 23:4 (1980): 369-376.

---, ed. Expert Systems in the Micro-Electronic Age. Edinburgh: Edinburgh University Press, 1979.

Miller, G.A.; Galanter, E.; and Pribram, K.H. Plans and the Structure of Behavior. New York: Holt, Rinehart and Winston, 1960.

Minksky, Marvin. Computation: Finite and Infinite Machines. Englewood Cliffs, NJ: Prentice-Hall, 1967.

---. "A Framework for Representing Knowledge." In The Psychology of Computer Vision, pp. 211-277. Edited by P.H. Winston. New York: McGraw-Hill, 1975.

---. Semantic Information Processing. Cambridge, MA: MIT Press, 1980.

Molnar, Richard. "Modelling Cognitive Constraints in Decision-Making: An Application to U.S. Foreign Policy." Paper presented to the annual meeting of the International Studies Association, Washington, DC, April 14-18, 1987.

Mueller, John E., ed. Approaches to Measurement in International Relations, 2d ed. New York: Appleton, Century, Crofts, 1973.

Mylopoulos, J. "An Overview of Knowledge Representation." SIGART Newsletter. 74 (January 1981): 5-12.

Nance, Richard E., and Tech, Virginia. "The Time and State Relations in Simulation Modeling." Communications of the ACM. 24:4 (April 1981): 173-179.

Naylor, Thomas H. Corporate Planning Models. Reading, MA: Addison-Wesley, 1979.

Naylor, Thomas H.; Balintfy, Joseph L.; Burdick, Donald S.; and Chu, Kong. Computer Simulation Techniques. New York: John Wiley and Sons, 1966.

Nelkin, Dorothy. The University and Military Research: Moral Politics at MIT. Ithaca, NY: Cornell University Press, 1972.

Newell, Alan; Shaw, J.C.; and Simon, Herbert A. "Chess-Playing Programs and the Problem of Complexity." IBM Journal of Research and Development. 2 (October 1958): 320-335.

Newell, Alan, and Simon, Herbert A. Human Problem Solving. Englewood Cliffs, NJ: Prentice-Hall, 1972.

Nii, H. Penny, and Aiello, Nelleke. "AGE (Attempt to Generalize): A Knowledge-Based Program for Building Knowledge-Based Programs." Proceedings of the sixth International Joint Conference on Artificial Intelligence, Tokyo, August 20-23, 1979, Vol. 2, pp. 645-655.

Nilsson, Nils J. Principles of Artificial Intelligence. Palo Alto, CA: Tioga Publishing Co., 1980.

Nixon, Richard M. The Real War. New York: Warner, 1980.

Oren, Tuncer I. "Concepts and Criteria to Assess the Acceptability of Simulation Studies: A Frame of Reference." Communications of the ACM. 24:4 (1981): 180-189.

Oren, Tuncer I., and Ziegler, Bernard P. "Concepts for Advanced Simulation Methodologies." Simulation. 32:3 (March 1979): 69-82.

O'Toole, Richard, and O'Toole, Anita W. "Negotiating Interorganizational Orders." The Sociological Quarterly. 22 (Winter 1981): 29-41.

Pease, Marshall III. "ACS.1: An Experimental Automated Command Support System." IEEE Transactions. Vol. SMC, 8:10 (October 1978): 726-735.

Peck, Merton J., and Scherer, Frederic M. The Weapons Acquisition Process: An Economic Analysis. Cambridge, MA: Harvard University Press, 1962.

Pierce, D.A. "Relationships--and the Lack Thereof--Between Economic Time Series, with Special Reference to Money, Reserves, and Interest Rates." Journal of the American Statistical Association. 72 (1977): 11-22.

Plutarch. Lives of Noble Grecians and Romans. Translated by John Dryden. Chicago: Encyclopaedia Britannica, 1952.

Pople, H.E.; Myers, J.D.; and Miller, R.A. "DIALOG: A Model of Diagnostic Logic for Internal Medicine." Proceedings of International Joint Conference on Artificial Intelligence, Cambridge, MA, August 22-25, 1977.

Preparata, F.P., and Yeh, R.T. Introduction to Discrete Structures. Reading, MA: Addison-Wesley, 1973.

Pressman, Jeffrey L., and Wildavsky, Aaron B. *Implementation: How Great Expectations in Washington Are Dashed in Oakland, Or, Why It's Amazing That Federal Programs Work at All*. Berkeley: University of California Press, 1973.

Pruitt, D.G., and Snyder, Richard C., eds. *Theory and Research on the Causes of War*. Englewood Cliffs, NJ: Prentice-Hall, 1969.

Raphael, B. *The Thinking Machine: Mind Inside Matter*. San Francisco: Freeman, 1976.

Rapoport, Anatol, and Chummah, Albert M. *Prisoner's Dilemma: A Study in Conflict and Cooperation*. Ann Arbor: University of Michigan Press, 1965.

Riker, William H., and Ordeshook, Peter C. *An Introduction to Positive Political Theory*. Englewood Cliffs, NJ: Prentice-Hall, 1973.

Roberts, Chalmers M. *The Nuclear Years: The Arms Race and Arms Control, 1945-1970*. New York: Praeger, 1970.

Roethlisberger, F.J., and Dickson, W. *Management and the Worker*. Cambridge, MA: Harvard University Press, 1939.

Rosenau, James N. "Puzzlement in Foreign Policy." *Jerusalem Journal of International Relations*. 1:4 (Summer 1976): 1-10.

---. "Thinking Theory Thoroughly." Address to the Seminar on International Relations Theory, Jawaharlal Nehru University, New Delhi, May 14-17, 1979.

---. *The Study of Political Adaptation*. London: Frances Pinter, 1981.

---. "The Microelectronic Revolution and the Conduct of Foreign Policy." Paper presented to the annual meeting of the American Political Science Association, Washington, DC, August 30 to September 2, 1984.

---, ed. *International Politics and Foreign Policy*, 2d ed. New York: Free Press, 1969.

---, ed. *In Search of Global Patterns*. New York: Free Press, 1976.

Rosenberg, David. "American Atomic Strategy and the Hydrogen Bomb Decision." *Journal of American History*. 66:2 (June 1979): 62-87.

---. "The Origins of Overkill: Nuclear Weapons and American Strategy, 1945-1960." *International Security*. 7:4 (Spring 1983): 3-71.

Rosenberg, Stephen. "Reasoning in Incomplete Domains." *Proceedings* of the sixth International Joint Conference on Artificial Intelligence, Tokyo, August 20-23, 1979, Vol. 2, pp. 735-737.

Rosenblueth, Arturo; Wiener, Norbert; and Bigelow, Julian. "Behavior, Purpose and Teleology." *Philosophy of Science*. 10 (1943): 18-24.

Rourke, Francis E. *Bureaucratic Power in National Politics*. Boston: Little, Brown and Co., 1972.

Sacerdoti, Earl. "The Nonlinear Nature of Plans." In *Advance Papers* of the fourth International Joint Conference on Artificial Intelligence. Cambridge, MA: MIT Artificial Intelligence Laboratory, 1975.

---. *A Structure for Plans and Behavior*. New York: Elsevier, 1977.

---. "Problem Solving Tactics." Proceedings of the sixth International Joint Conference on Artificial Intelligence, Tokyo, August 20-23, 1979, Vol. 2, pp. 1077-1085.

Salmon, Trevor C. "Rationality and Politics: The Case of Strategic Theory." British Journal of International Affairs. 2:2 (October 1976).

Salomon, M.D. "New Concepts for Strategic Parity." Survival. 19:6 (November/December 1977): 255-262.

Sanchez, D.A. Ordinary Differential Equations and Stability Theory: An Introduction. San Francisco: W.H. Freeman, 1968.

Sapolsky, H.M. The Polaris System Development: Bureaucratic and Programmatic Success in Government. Cambridge, England: Cambridge University Press, 1972.

Schank, Roger, and Colby, Kenneth M. Computer Models of Thought and Language. San Francisco: Freeman, 1973.

Schelling, Thomas. The Strategy of Conflict. London: Oxford University Press, 1960.

---. Arms and Influence. New Haven: Yale University Press, 1966.

Schrodt, Philip A. "Pattern Recognition and the Analysis of International Event Sequences in the Middle East." Paper presented to the annual meeting of the International Studies Association, Washington, DC, March 5-9, 1985.

---. "Data Structures for an Artificial Intelligence Model of Foreign Policy Decision-Making." Paper presented to the annual meeting of the International Studies Association, Anaheim, CA, March 25-29, 1986.

---. "Classification of Interstate Conflict Outcomes Using a Bootstrapped CLS Algorith." Paper presented to the annual meeting of the International Studies Association, Washington, DC, April 14-18, 1987.

---. "Short Term Predictions of International Events Using Pattern Matching." Paper presented to the annual meeting of the International Studies Association, Washington DC, April 14-18, 1987.

---. "Machine Learning Techniques Applied to the Analysis of International Events." Paper presented to the annual meeting of the American Political Science Association, Chicago, September 3-6, 1987.

Schumpeter, Joseph. Imperialism and Social Classes: Two Essays. New York: Signet, 1951.

Schutz, Alfred. The Phenomenology of the Social World. Translated by George Walsh and Frederick Lehnert. Evanston, IL: Northwestern University Press, 1967.

Searle, John R. "Minds, Brains, and Programs." In Mind Design: Philosophy, Psychology, Artificial Intelligence, pp. 282-306. Edited by John Haugeland. Cambridge, MA: MIT Press, 1981.

Sherry, M.S. Preparing for the Next War: America Plans for Post-war Defense 1941-1945. New Haven: Yale University Press, 1977.

Shortliffe, E.H. Computer-Based Medical Consultation: MYCIN. New York: American Elsevier, 1976.

Shortliffe, E.H.; Buchanan, B.G.; and Feigenbaum, E.A. "Knowledge Engineering for Medical Decision-Making: A Review of Computer-based Clinical Decision Aids." Proceedings of the IEEE. 67 (1979): 1207-1224.

Simon, Herbert A. Models of Man. New York: John Wiley and Sons, 1957.

---. Adminstrative Behavior: A Study of the Decision-Making Process in the Administrative Organization. New York: Free Press, 1965.

---. The Sciences of the Artificial. Cambridge, MA: MIT Press, 1969, 1981.

---. "Artificial Intelligence Research Strategies in the Light of AI Models of Scientific Discovery." Proceedings of the sixth International Joint Conference on Artificial Intelligence, Tokyo, August 20-23, 1979, Vol. 2, pp. 1086-1094.

---. "Human Nature in Politics: The Dialogue of Psychology with Political Science." American Political Science Review. 79:2 (June 1985): 293-304.

Smith, Perry M. The Air Force Plans for Peace 1943-1945. Baltimore: Johns Hopkins Press, 1970.

Smith, Reid G. "The Contract Net: A Formalism for the Control of Distributed Problem Solving." Proceedings of the fifth International Joint Conference on Artificial Intelligence, Cambridge, MA, August 22-25, 1977, Vol. 2, p. 472.

---. "A Framework for Distributed Problem Solving." Proceedings of the sixth International Joint Conference on Artificial Intelligence, Tokyo, August 20-23, 1979, Vol. 2, pp. 836-841.

Smith, Reid G., and Baker, James D. "The DIPMETER Advisor System: A Case Study in Commercial Expert System Development." Proceedings of the eighth International Joint Conference on Artificial Intelligence, Karlsruhe, Germany, August 8-12, 1983, Vol. 1, pp. 123-129.

Snow, C.P. Science and Government. Cambridge, MA: Harvard University Press, 1961.

Snyder, Glenn H., and Diesing, Paul. Conflict among Nations: Bargaining, Decision Making and System Structure in International Crises. Princeton, NJ: Princeton University Press, 1974.

Snyder, R.C.; Bruck, H.W.; and Sapin, B. Foreign Policy Decision-Making. New York: Free Press, 1962.

Spielmann, K. Analyzing Soviet Strategic Arms Decisions. Boulder, CO: Westview Press, 1978.

Sprout, Margaret Tutle. "Mahan: Evangelist of Sea Power." In Makers of Modern Strategy: Military Thought from Machiavelli to Hitler, pp. 416-445. Edited by E.M. Earle. Princeton, NJ: Princeton University Press, 1971.

Steele, Guy L., Jr. Common LISP. Bedford, MA: Digital Press, 1984.

Steinbruner, John D. The Cybernetic Theory of Decision: New Dimensions of Political Analysis. Princeton, NJ: Princeton University Press, 1974.

Steinbruner, John D., and Carter, Barry. "Organizational and Political Dimensions of the Strategic Posture: The Problems of Reform." Daedalus. 104 (Summer 1975): 131-154.

Stogdill, R.M. The Process of Model-Building in the Behavioral Sciences. Columbus: Ohio State University Press, 1970.

Strauss, Anselm. Negotiations: Varieties, Contexts, Processes and Social Order. San Francisco: Jossey-Bass Publishers, 1978.

Strauss, B.S. "Thucydides on the Insecurity of Tyranny: A Comment on Professors Lebow and Bluhm." PS. 17 (1984): 594-596.

Sugarman, Robert, and Wallich, Paul. "The Limits to Simulation." IEEE Spectrum. 20:4 (April 1983): 36-41.

Sun Tzu. The Art of War. Translated by D. Griffith. London: Oxford University Press, 1963.

Sylvan, David, and Majeski, Stephen. "Multiple Actor Advisory Simulations in Foreign Policy." Paper presented to the annual meeting of the International Studies Association, Anaheim, CA, March 25-29, 1986.

Sylvan, Donald A., and Chan, Steve, eds. Foreign Policy Decision Making: Perception, Cognition, and Artificial Intelligence. New York: Praeger, 1984.

Tamashiro, Howard. "Algorithms, Heuristics, and Artificial Intelligence Modelling of Strategic Statecraft." In Foreign Policy Decision Making: Perception, Cognition, and Artificial Intelligence, pp. 197-226. Edited by Donald A. Sylvan and Steve Chan. New York: Praeger, 1984.

Tanaka, Akihiko. "China, China Watching, and CHINA_WATCHER." In Foreign Policy Decision Making: Perception, Cognition, and Artificial Intelligence, pp. 310-344. Edited by Donald A. Sylvan and Steve Chan. New York: Praeger, 1984.

Tarr, David W. "Military Technology and the Policy Process." Western Political Quarterly. 18 (March 1965): 135-148.

Taylor, Phillip. "Do Technologies Determine Strategies? Trends in Naval Strategies and Tactics." Paper presented to the annual meeting of the International Studies Association, Atlanta, March 27 to April 1, 1984.

Teitelman, W. The INTERLISP Reference Manual. Palo Alto, CA: Xerox Palo Alto Research Center, 1974.

Thiel, H. Principles of Econometrics. New York: Wiley, 1971.

Thompson, W.R. "Comparing World Systems: Systemic Leadership Succession and the Peloponnesian War Case." Paper presented to the annual meeting of the International Studies Association, Anaheim, CA, March 25-29, 1986.

Thorson, Stuart J. "Intentional Inferencing in Foreign Policy: An AI Approach." In Foreign Policy Decision Making: Perception, Cognition, and Artificial Intelligence, pp. 280-309. Edited by Donald A. Sylvan and Steve Chan. New York: Praeger, 1984.

---. "The Intentional Inferencing Problem in Foreign Policy Decision Making." Paper presented to the annual meeting of the International Studies Association, Washington, DC, March 5-9, 1985.

Thorson, Stuart J., and Andersen, Kristi. "A Conceptual Model of Intentional Inferencing in Foreign Policy." Paper presented to the annual meeting of the International Studies Association, Anaheim, CA, March 25-29, 1986.

Thorson, Stuart J., and Bennett, James. "The Use of Rules in Computational Modelling of International Politics." Paper presented to the annual meeting of the International Studies Association, Anaheim, CA, March 25-29, 1986.

Thorson, Stuart J., and Sylvan, Donald A. "Counterfactuals and the Cuban Missile Crisis." International Studies Quarterly. 26:4 (December 1982): 539-571.

Thucydides. The Peloponnesian War. Translated by F. Crawley. New York: Random House, 1951.

Timberlake, Clare H. "Planning of U.S. Arms Policy." In The Prospects for Arms Control, pp. 145-151. Edited by James E. Dougherty and John F. Lehman, Jr. New York: MacFadden-Bartel, 1965.

Tuchman, Barbara. The Guns of August. New York: Dell, 1962.

United States. Defense Nuclear Agency. Measures and Trends: U.S. and U.S.S.R. Strategic Force Effectiveness. Washington, DC: Santa Fe Corp., 1978.

Van Creveld, Martin. Command in War. Cambridge, MA: Harvard University Press, 1985.

Veale, F.J.P. Advance to Barbarism: The Development of Total Warfare from Serajevo to Hiroshima. New York: Devon-Adair, 1968.

Von Clausewitz, Karl. On War. Translated by Anatol Rapoport. Princeton, NJ: Princeton University Press, 1976.

Von Leibnitz, Georg Wilhelm F. The Monadologie. Translated by H.W. Carr. Los Angeles: University of Southern California, 1930.

Von Neumann, John. The Theory of Self-Reproducing Automatia. Urbana: University of Illinois Press, 1966.

Von Neuman, John, and Morgenstern, Oscar. "Zur Theorie der Gesellschaftspiele." Mathematische Annalen. 100 (1928): 295-320.

---. Theory of Games and Economic Behavior. Princeton, NJ: Princeton University Press, 1944.

Von Wright, Georg Henrik. Explanation and Understanding. Ithaca, NY: Cornell University Press, 1971.

Waterman, D., and Hayes-Roth, F. Pattern-Directed Inference Systems. New York: Academic Press, 1979.

Weber, Max. The Theory of Social and Economic Organization. Edited by Talcott Parsons. Translated by A.M. Henderson and T. Parsons. New York: Oxford University Press, 1947.

---. Economy and Society. Translated by A.M. Henderson and T. Parsons. Glencoe, IL: Free Press, 1957.

---. "The Interpretive Understanding of Social Action." In Readings in the Philosophy of the Social Sciences, pp. 19-33. Edited by May Brodbeck. New York: Macmillan, 1968.

Weissman, C. LISP 1.5 Primer. Belmont, CA: Dickenson, 1967.

Weizenbaum, J. "ELIZA--A Computer Program for the Study of Natural Language Communication between Man and Machine." Communications of the ACM. 9 (1966): 36-45.

Weltman, John J. Systems Theory in International Relations. Lexington, MA: D.C. Heath, 1973.

Wesson, Robert G. "Non-spending in Secret for Defense: Big Budgets Impress the Soviets, So Dummy Outlays Could Help." Los Angeles Times, February 15, 1985.

Wiener, Norbert. The Human Use of Human Beings: Cybernetics and Society. Boston: Houghton-Mifflin Co., 1950.

---. Cybernetics, or Control and Communication in the Animal and the Machine, 2d ed. Cambridge, MA: MIT Press, 1961.

---. God and Golem, Inc.: A Comment on Certain Points Where Cybernetics Impinges on Religion. Cambridge, MA: MIT Press, 1964.

Wildavsky, Aaron. The Politics of the Budgetary Process. Boston: Little, Brown and Co., 1964.

Winch, Peter. The Idea of a Social Science. London: Routledge and Kegan Paul, 1958.

Winograd, Terry. "Frame Representation and the Declarative-Procedural Controversy." In Representation and Understanding, pp. 185-210. Edited by Daniel Bobrow and A. Collins. New York: Academic Press, 1975.

Winston, Patrick H. Artificial Intelligence. Reading, MA: Addison-Wesley, 1977.

Winston, Patrick H., and Horn, B.K.P. LISP. Reading, MA: Addison-Wesley, 1981.

Wirth, Niklaus. Algorithms and Data Structures. Englewood Cliffs, NJ: Prentice-Hall, 1986.

Woodward, E.L. Great Britain and the German Navy. Oxford, England: Claredon Press, 1935.

Yu, V.L.; Fagan, L.M.; Wraith, S.M.; Clancey, W.J.; Scott, A.C.; Hannigan, J.F.; Blum, R.L.; Buchanan, B.G.; and Cohen, S.N. "Antimicrobal Selection by Computer: A Blinded Evaluation by Infectious Disease Experts." Journal of the American Medical Association. 242 (1979): 1279-1282.

Zinnes, Dina A. "Three Puzzles in Search of a Researcher." International Studies Quarterly. 24:1 (March 1980): 1-27.

Zisman, Michael D. "Representation, Specification, and Automation of Office Procedures." Ph.D. dissertation, University of Pennsylvania, Wharton School Department of Decision Sciences, 1977.

INDEX

About the Author

CRAIG ETCHESON is a Research Associate at the Institute for Transnational Studies at the University of Southern California where he teaches research methodology and international relations theory. He is currently working on two collaborative projects, one on models of international arms transfers, and another on revolutionary politics in Peru. Etcheson is Manager of Financial Systems for Paramount Pictures Corporation. His previous publications include *The Rise and Demise of Democratic Kampuchea* and "Applying AI Techniques to the Study of War."